Wirtschaft + Gesellschaft

Reihe herausgegeben von

Andrea Maurer, Fachbereich IV Soziologie, Universität Trier, Trier, Germany

Mark Lutter, Bergische Universität Wuppertal, Wuppertal, Nordrhein-Westfalen, Germany

Wirtschaft und Gesellschaft ist ein wichtiges Themenfeld der Sozialwissenschaften. Daher diese Buchreihe: Sie will zentrale Institutionen des Wirtschaftslebens wie Märkte, Geld und Unternehmen sowie deren Entwicklungsdynamiken sozial- und gesellschaftstheoretisch in den Blick nehmen. Damit soll ein sichtbarer Raum für Arbeiten geschaffen werden, die die Wirtschaft in ihrer gesellschaftlichen Einbettung betrachten oder aber soziale Effekte des Wirtschaftsgeschehens und wirtschaftlichen Denkens analysieren. Die Reihe steht für einen disziplinären wie theoretischen Pluralismus und pflegt ein offenes Themenspektrum.

Lara Maria Espeter

From Hobby to Billion-Dollar Industry

Mapping the Commodity Chain of Cut Flowers in the Longue Durée

Springer VS

Lara Maria Espeter
Technische Universität Berlin
Berlin, Germany

Zugl. Berlin, Technische Universität, Diss., 2025 u. d. T. "The Commodity Chain of Cut Flowers in the Longue Durée. Long-Term Processes of Establishing and Maintaining Power in the World-Economy"

ISSN 2626-6156 ISSN 2626-6164 (electronic)
Wirtschaft + Gesellschaft
ISBN 978-3-658-52436-4 ISBN 978-3-658-52437-1 (eBook)
https://doi.org/10.1007/978-3-658-52437-1

This Springer VS imprint is published by the registered company Springer Fachmedien Wiesbaden GmbH, part of Springer Nature.
The registered company address is: Abraham-Lincoln-Str. 46, 65189 Wiesbaden, Germany

To S, I and V.

Acknowledgement I acknowledge support by the Open Access Publication Fund of TU Berlin and the Deutsche Forschungsgemeinschaft (DFG, German Research Foundation) under Projectnumer 290045248 (SFB 1265) and Projectnumber 458460698 (Apples and Flowers).

Competing Interests The author has no competing interests to declare that are relevant to the content of this manuscript.

Abstracts

English
This book examines the process-nature of economic structures as commodity chains within the modern world-system. Building on the theoretical and methodological concepts developed by Hopkins and Wallerstein, it introduces 'process-oriented commodity chain mapping' as a method for analysing and understanding the process-nature of economic structures. The cut flower commodity chain, with a particular focus on the Aalsmeer region, serves as a case study to illustrate the application of this methodological concept. The analysis encompasses the structure of actors, the spatial structure, and the inherent power structure within content-based periods from the 1850s to 2022. Furthermore, the book analyse the process-nature of the commodity chain within and across these periods, enabling the mapping of power establishment and maintenance processes within the cut flower commodity chain. By demonstrating the practical implementation of process-oriented commodity chain research, this book makes an empirical contribution to world-system analysis by providing a concrete definition of power within economic structures and introducing the collective actors in the modern world-system.

Deutsch
In diesem Buch wird der prozessuale Charakter wirtschaftlicher Strukturen als Warenkette innerhalb des modernen Welt-Systems untersucht. Aufbauend auf den von Hopkins und Wallerstein entwickelten theoretischen und methodologischen Konzepten wird das „Process-Oriented Commodity Chain Mapping" als Erhebungs- und Analysemethode für den prozessualen Charakter wirtschaftlicher Strukturen eingeführt. Die Warenlieferkette für Schnittblumen mit besonderem

Fokus auf die Region Aalsmeer dient als Fallstudie, um die Anwendung dieses methodischen Konzepts zu veranschaulichen. Die Analyse umfasst Dabei werden der Akteure, die räumliche Struktur und die inhärente Machtstruktur zwischen 1850 und 2022. Das Buch analysiert darüber hinaus den prozessualen Charakter der Warenkette innerhalb und über inhaltsgeleitete Perioden hinweg. Dadurch wird die Darstellung von Prozessen der Machtetablierung und -erhaltung innerhalb der Schnittblumenlieferkette möglich. Durch die Darstellung der praktischen Umsetzung prozessorientierter Warenkettenforschung leistet dieses Buch ein empirischen empirischen Beitrag zur Welt-System-Analyse, indem es diese um eine konkrete Machtdefinition in wirtschaftlichen Strukturen ergänzt und kollektive Akteure im modernen Welt-System einführt.

Contents

List of Figures

1 From Hobby to Billion-Dollar Industry

'We shall try to draw as detailed a picture as possible of [...] how the commodity chain functioned. We do this, to be sure, in order to demonstrate the complexity of functioning of such a commodity chain, but we do this as well to make the case that what we see then is not strikingly different from what we see today. We hope thereby to illuminate the fundamental processes of a capitalist world-economy' (Wallerstein 2000a, 3).

The history of cut flowers as a commodity is not merely the story of a fragile, perishable consumer product. Rather, it connects research on the concept of heredity, imperialism, and the perception of the Netherlands from an external perspective. The nation-state Netherlands and flowers are inextricably linked. What began as a hobby for a group of upper-middle-class individuals has, over the last 400 years, developed into a global industry with a turnover of billions. How did this happen? And why, when dealing with cut flowers, is it impossible to avoid the Netherlands or its people? What is it about the flower auction in Aalsmeer that makes it such an important intermediary in the commodity chain?

In this book, using the cut flower industry and its centre, the flower auction in Aalsmeer[1], as a case study in the longue durée, I examine the process-nature of economic structures.

Considering the backdrop of multiple external shocks (Follmann et al. 2024; Espeter and Hering 2025), the stability of global economic structures is particularly notable. This raises the question of which factors are responsible for the longevity of these structures, and which processes lead to continuity or change. To explore the process-nature of economic structures, this study will examine them in the longue durée.

---

[1] In the following, both the location and the actor 'flower auction' are referred to synonymously.

L. M. Espeter, *From Hobby to Billion-Dollar Industry*, Wirtschaft + Gesellschaft,
https://doi.org/10.1007/978-3-658-52437-1_1

The structure of today's economy can only be understood by taking its history into account. The economy would resemble something very different if it had not undergone a wide variety of historical processes. Marx and Engels described the process-nature of social developments as early as 1849 in the 'Communist Manifesto' (2015 [1848]). They trace the emergence of modern society by examining the social orders that preceded it. Similarly, Max Weber regarded society as something that had grown historically, 'becoming what it is and not something else' (Weber 2002 [1904]]).

Examining the modern world-system in his work, 'The Modern World-System I–IV' (2011a [1976]; 2011b [1980]; 1989; 2011c), Immanuel Wallerstein is one of the co-founders of world-system analysis and remains one of its most prominent scholars. As part of the neo-Marxist school of thought, world-systems analysis examines society as a structure that has evolved historically and is undergoing change, inevitably transitioning into a different form of society. This is because: 'Structures are those coral reefs of human relations which have a stable existence over relatively long periods of time. But structures too are born, develop, and die' (Wallerstein 2011a, 3 [1976]).

Methodologically, world-systems analysis differs from national perspectives in that it primarily uses the modern world-system as its unit of analysis. The focus is on the capitalist world-economy that emerged in a historic process and the axial division of labour within it. This arrangement benefits the core financially at the periphery's expense. Consequently, world-systems analysis is an extremely suitable concept for examining the process-nature of economic structures. Therefore, I examine commodity chains as structuring elements of the capitalist world-economy (Chase-Dunn 1998; Hopkins and Wallerstein 1977; 1994a; 2016). Commodity chain research has its origins in world-systems analysis and is particularly suited to investigating process-nature in the longue durée. Hopkins and Wallerstein (1977) originally introduced the concept of the commodity chain for this purpose. They were most likely the first scholars since Adam Smith (1999 [1776]) to define a commodity chain, while also developing a methodological concept and future research questions for examining commodity chains in the longue durée.

To date, research on commodity chains has developed into a broad multi- and interdisciplinary field, with various theoretical strands emerging. These include global value chains (Gereffi, Humphrey, and Sturgeon 2005) and global production networks (Coe, Dicken, and Hess 2008), with research spanning the academic fields of economics, geography, political economy, sociology, and others (see f.e. Bargawi and Newman 2017; Bair et al. 2021; Alford et al. 2025; Butollo et al. 2025).

Conducting commodity chain research from a world-systems analytical perspective means that the resulting empirical work provides valuable insights into economic structures and their process-nature, since consideration commodity chains in the longue durée is an integral part of the approach. Nevertheless, that there is a gap in research: While some works within the framework of world-system analysis consider economic structures in the modern world-system, they tend to omit the concept of commodity chains in the longue durée. Reilly (1993), for example, examines the impact of the integration of craft businesses in the Ottoman Empire into the world-economy. Moore (2000) analyses the spatial expansion of the world-economy and its impact on nature. Applying the commodity chain approach would reveal the complex economic interrelationships that are missed when this approach is not used. Other works examine commodity chains in the modern world-system in the longue durée, but use different theoretical approaches to those of Hopkins and Wallerstein. Examples include the works of Talbot (2004; 2011), Quark (2014), Vieira (2014), and Korzeniewicz and Payne (2019). Thereby, they address a wide range of issues, but do not explicitly discuss the process-nature of economic structures in the longue durée or conduct comparable studies.

After extensive literature research, I conclude that Özveren (1994; 2000) and Pelizzon (1994; 2000) are the only scholars who empirically examine commodity chains within the modern world-system, based on the theoretical and methodological framework of Hopkins and Wallerstein. The studies of Özveren and Pelizzon are connected, serving as contrasting examples and demonstrating different approaches to analysing commodity chains in the longue durée. Their period under study covers the years 1590–1790. Through their work, they reveal the process-nature of economic structures, which are influenced by internal and external processes at various stages of the commodity chain. They also demonstrate the interplay of these processes by considering the longue durée.

As previously mentioned, commodity chains are researched by a wide variety of disciplines. In particular, these is extremely intriguing work of several historians. Their research examines the longue durée, focusing on an industry or product, and is mostly conducted without a specific theoretical framework. This is relevant to my study insofar as there are similarities between their empirical work and my own. To illustrate this point, I will refer to Nadri (2016), Sapoznik (2016), and Baraibar Norberg and Deutsch (2023).

In his work from 2016, the historian Nadri examines the production and trade of indigo in South Asia and India between 1580 and 1930. The study covers this 350-year period using a commodity chain approach, which allows him to examine the nature of economic processes. The time under study is divided into content-based sub-periods (Nadri 2016). The historian Sapoznik (2019) analyses the medieval economy (1300–1550) in the context of the commodity beeswax. Based on historical data, she examines the influence of consumption, particularly for religious purposes, and its impact on production and trade. To this end, she examines import volumes and fluctuations in wax prices over a long period of time. In their 2023 study, the historians Baraibar Norberg and Deutsch examine a commodity over a period of almost 2,000 years. Their work builds on Gereffi's approach. However, they approach this subject from the perspective of political economy within the framework of world-systems analysis. For their analysis in the longue durée, they first divide the period under investigation into hegemon-based periods, demonstrating a connection to Arrighi's work. Furthermore, Baraibar Norberg and Deutsch provide a detailed description of the difficulties they experienced when collecting and analysing the data (Baraibar Norberg and Deutsch 2023).

Nadri (2016), Sapoznik (2016) and Baraibar, Norberg and Deutsch (2023) demonstrate that historical analysis of economic structures conducted by historians provides a comprehensive understanding of their development and evolution. However, what limits the traceability and comparability of their data and results is that, with the exception of Baraibar Norberg and Deutsch (2023), they do not explicitly state their methodological approach. The process of reconstructing the commodity chain and the factors influencing the periodisation, case selection, data collection and analysis remain unclear. As mentioned above, historians have traditionally led research into commodity chains in the longue durée. In this book, I will demonstrate the benefits of adopting a sociological approach. The aim of sociology is to identify and generalise patterns, regularities, structures and processes. This enables social phenomena to be analysed within a theoretical framework and explanations to be found for them. Furthermore, sociology, and thus my work, places a more explicit focus on a methodological concept. This raises fundamental questions about generalisability and the understanding of cause-and-effect relationships. While historical science sheds light on the depth of individual events, historical sociology, with its analytical and generalising approach, can contribute to a more comprehensive understanding of complex social developments.

The central research question of this book concerns the process-nature of economic structures, divided into a methodological concept and an empirical study. The study I examine in this book is rooted in world-systems analysis and the theoretical and methodological concept of commodity chains, as developed by Hopkins and Wallerstein. I outline this framework in Chap. 2.

In Chap. 3, I address the methodological question of how the economic structures of the modern world-system and their process-nature can be identified and analysed. To this end, I introduce process-oriented commodity chain mapping. The aim is to provide an approach that will be useful for future research on this topic. I therefore integrate process-oriented methods (Baur 2017; Hergesell, Baur, and Braunisch 2020) with case study approaches (Yin 2018). I distance my approach from previous periodisation schemes and advocate for the identification of content-based, so-called meaningful periods. Furthermore, I supplement the theoretical concepts of commodity chains and the modern world-system with an explicit definition and operationalisation of power within commodity chains. This enables me to identify the inherent power structure within the commodity chain.

Before empirically examining the research question, concerning the process-nature of economic structures, I will implement the methodology in Chap. 4. Building on a detailed reconstruction of the commodity chain for Kenyan cut flowers (Espeter 2025), I focus on the historical development of the cut flower commodity chain, paying particular attention to the region around Aalsmeer. This case is particularly well suited to examining the process-nature of economic structures, given the emergence of a unique central actor—the cooperative flower auction in Aalsmeer—and the industry's notable spatial and organisational transformations in the longue durée.

My analysis traces how actors within commodity chains have adapted and repositioned themselves over time and space in response to changing conditions. Therefore, I reconstruct the cut flower commodity chain within five identified meaningful periods, considering the structure of actors, spatial structure and inherent power structure in Chap. 6. In the next analytical step, I examine the process-nature of this commodity chain within and across the meaningful periods. To improve the understanding of the text, this step is examined before the first step in Chap. 5. Focusing particularly on the establishment and maintenance of power within the cut flower commodity chain, the empirical analysis identifies the processes that have shaped the chain in the longue durée. Central here are the relationships between the state and firms, as well as monopolisation tendencies within the chain.

My study demonstrates the value of commodity chain analysis in the longue durée for achieving a thorough understanding of the structure and dynamics of the world-economy. I emphasise the advantages of a historical approach to examining economic structures, using the process-oriented commodity chain mapping method that I introduce. In this way, I demonstrate how such a reconstruction of economic structures can be carried out methodologically, and that it is beneficial for gaining a comprehensive understanding of economics and the processes involved.

This book contributes to the theoretical discourse on commodity chains and the modern world-system. It reveals that the establishment and maintenance of power, as well as the relationship between state actors and companies, have significantly shaped the commodity chain. Contrary to some theoretical expectations, state protectionism did not result in quasi-monopolies for individual companies, but rather promoted the development of the entire sector. Power dynamics and various internal and external factors have influenced the development of the cut flower commodity chain from its origins to the present day. Furthermore, I demonstrate that the spatial expansion of the capitalist world-economy can occur through the acquisition of commercially usable land, the expansion of sales markets and access to cheap labour-power within the region under investigation.

Part I

Theory and Methods

Commodity Chains in the Modern World-System

2

'I argue here two things: One, the items that are meant to be consumed directly (let us designate them as finished products) are in fact the outcome of a long series of production processes we shall call a commodity chain, and that such commodity chains typically are geographically extensive and contain many kinds of production units within them with multiple modes of remunerating labor. Two, such trans state, geographically extensive, commodity chains are not a recent phenomenon, dating from say the 1970's or even 1945, but that they have been an integral part, and a major part, of the functioning of the capitalist world-economy since it came into existence' (Wallerstein 2000a, 2).

In the following, I will outline the theoretical framework of this book. It is rooted in the work of Immanuel Wallerstein, one of the founders of world-systems analysis and a prominent representative of this approach. Alongside Terence Hopkins, he is also a pioneer in the field of commodity chain research.

In this book, I build on the theoretical and methodological considerations of Wallerstein and Hopkins. My research question concerning the process-nature of economic structures is rooted in Wallerstein's work of the modern world-system in the longue durée (2011a [1976]; 2011b [1980]; 1989; 2011c).

In this chapter, I will therefore (1) outline the theoretical concept of world-system analysis as described by Wallerstein. This approach is extremely comprehensive, so I will focus on the aspects that are relevant to this book. To this end, I (2) describe the basic features of Wallerstein's world-systems analysis that are crucial to this book: world-economy, axial division of labour, states and firms as institutions of the modern world-systems, visualised in Fig. 1 and followed by (3) the relationship between states and companies within it. After elaborating on the concept of commodity chains (4), I will conclude by (5) presenting the methodological considerations for collecting and analysing these chains according

L. M. Espeter, *From Hobby to Billion-Dollar Industry*, Wirtschaft + Gesellschaft,
https://doi.org/10.1007/978-3-658-52437-1_2

to Hopkins and Wallerstein, as well as (6) the current state of research referring to commodity chains in the longue durée.

2.1 World-Systems Analysis

The aim of the world-systems analysis is to capture social reality. All world-systems, such as the modern world-system, have emerged through a historical process. Within this theoretical framework, I examine the process-nature of economic structures.

Wallerstein's major work 'The Modern World-System I-IV' (2011a [1976]; 2011b [1980]; 1989; 2011c) established him as one of the founders of world-systems analysis. Alongside Samīr Amīn (Amīn 1974) and André Gunder Frank (Frank 1978), he remains its most influential scholar until today. The theoretical foundations of world-systems analysis are largely based on the work of Karl Marx (Marx 1887 [1867]; Marx and Engels 2015 [1848]) and Fernand Braudel (Braudel 1996a [1949]; 1996b [1949]), as well as on the dependency theory developed in Latin America in the 1960 s (Frank 1978).

Wallerstein formulates the goal of world-systems analysis as explaining long-term social change. The understanding of time is influenced by Braudel's concept of the longue durée. The longue durée refers to a social time in which the emergence and change of overall social structures are to be understood as historical processes (Braudel 1996a [1949]).

They emerge through a historical process and have a life that begins, evolves and ends. Each world-system is preceded by another and arises out of the reaction to the crisis of the previous world-system (Wallerstein 2006, 1). All world-systems are therefore inherently processual.

World-systems analysis replaces the nation-state within its political borders with the world-system as the unit of analysis at the centre of its consideration (Wallerstein 2006, 16). Wallerstein distinguishes two types of world-systems: hegemony and world-economy. Wallerstein defines a hegemony as a world-system dominated by a single political power. This power defines the system for a certain period of time, but cannot secure its position indefinitely (Wallerstein 2006, 57–59). Examples of this are the hegemony of the United Provinces (~1650), Great Britain (~1850) and the USA (~1950) (Wallerstein 2006, 57; Arrighi 1994). States and regions within a world-economy form an economic unit that influence and hold the whole system together. The relations between regions in this type of world-system are primarily economic (Wallerstein 2011a [1976], 15). They are therefore defined by their trade and production processes.

2.2 The Modern World-System

Wallerstein, like Marx and Engels before him (Marx and Engels 2015 [1848]), identifies the crisis of feudalism in the sixteenth century as the initial cause of the emergence of today's western society (Wallerstein 2011a [1976], 38). Feudalism describes the social, political and economic system that prevailed in Western and Central Europe since the Middle Ages. Monarchs and the nobility were the ruling class. By the time of the French Revolution in 1789, new forms of government had been established (Wallerstein 2011a [1976], 58–59). European states responded to the crisis of feudalism by, among other things, expanding their spheres of influence. In this process, other regions of the world were brought under the control of the respective state, mostly by force. Consequently, the first colonisations are the foundation of the modern world-system (Wallerstein 2011b [1980], 7–8). Access to new material and immaterial resources, such as land, raw materials and labour, made the economic structure we know today possible.

Therefore, the modern world-system, like any other world-system, does not necessarily include the whole globe. At the time of its emergence, the modern world-system initially encompassed parts of Europe and Latin America (the first colonised regions) and gradually extended to large parts of the world through the European expansion based on imperialism (Wallerstein 2006, 23).

As previously stated, the modern world-system is a world-economy. Concretely, the *capitalist* world-economy is inherent in the modern world-system (Wallerstein 2006, 23). Wallerstein states: '[w]e are in a capitalist system only when the system gives priority to the endless accumulation of capital. Using such a definition, only the modern world-system has been a capitalist system' (Wallerstein 2006, 24). Wallerstein defines the world-economy as based on monetary exchange (Wallerstein 2006, 17). In theory, these markets can be an entire free market—meaning they would be completely independent of external (political) influences. However, various protectionist measures create smaller markets within which products are not freely available for purchase or sale. '[A] totally free market [...] would make impossible the endless accumulation of capital' (Wallerstein 2006, 25). This would make it unattractive for capitalist endeavours. Because '[i]n such a perfect market, it would always be possible for the buyers to bargain down the sellers to an absolutely minuscule level of profit [...], and this low level of profit would make the capitalist game entirely uninteresting to producers, removing the basic social underpinnings of such a system' (Wallerstein 2006, 25–26).

The capitalist world-economy requires spatial expansion—whether material or financial (Arrighi 1994; Karataşli i.P.)—and hierarchisation within the system itself. The spatial expansion of the system makes cheap labour and natural resources, which were previously outside the scope of the modern world-system, accessible to it. This creates constant cheap production conditions and new sales markets for manufactured products. The territorial and hierarchical differentiation is achieved through the distinction core and periphery (Wallerstein 2011b [1980], 179) based on the axial division of labour. Cultural and political heterogeneity is inherent to the modern world-system. The identity-forming and structuring feature of the modern world-system is the axial division of labour. This enables the system to develop and ensures its continued existence (Wallerstein 2006, 24).

Based on Prebisch (1963) the axial division of labour refers to the spatial and functional division of production processes within the system. It affects the scope of action of the fundamental institutions of the modern world-system, such as states within the interstate system and firms operating in markets. Their position within the system, resulting from the axial division of labour, is reflected in their strength, power and influence. These institutions shape and sustain the system (Wallerstein 2006, 24–25). Therefore: Social reality and the resulting options for action in the modern world-system are determined by the system itself, as they can be traced back to one's position in the system (Wallerstein 2006, 1). The distinction and hierarchy of regions is based on the production processes located there and refers to the economic link between the core and the periphery.

The axial division of labour takes place formally through the type of production processes and spatially through their location. Regions in which highly monopolised production processes predominate are the core (Wallerstein 2006, 28). Monopolised production processes are able to generate higher added value. This is due to their unique position in the relationship between the state and firms. They generate high profits that increase the influence and strength of the core regions. Wallerstein uses the term quasi-monopoly (or relative monopoly) to emphasise that monopolies are ideal and are basically non-existent (Wallerstein 2006, 24). Wallerstein describes quasi-monopolies as a special form of productive firm characterised by a 'wide margin the costs of production and sales prices' and a 'high rates of profit' (Wallerstein 2006, 24). They only exist for a certain period of time. As soon as their market power declines, the capital accumulated through the monopoly position is transferred to another product or industry. For such quasi-monopolies to emerge and persist, firms need the protection of a strong state that can enforce their interests against other firms and states.

The periphery is identified by political weakness both inside and outside its political borders. Wages are low and the goods produced are mostly for daily

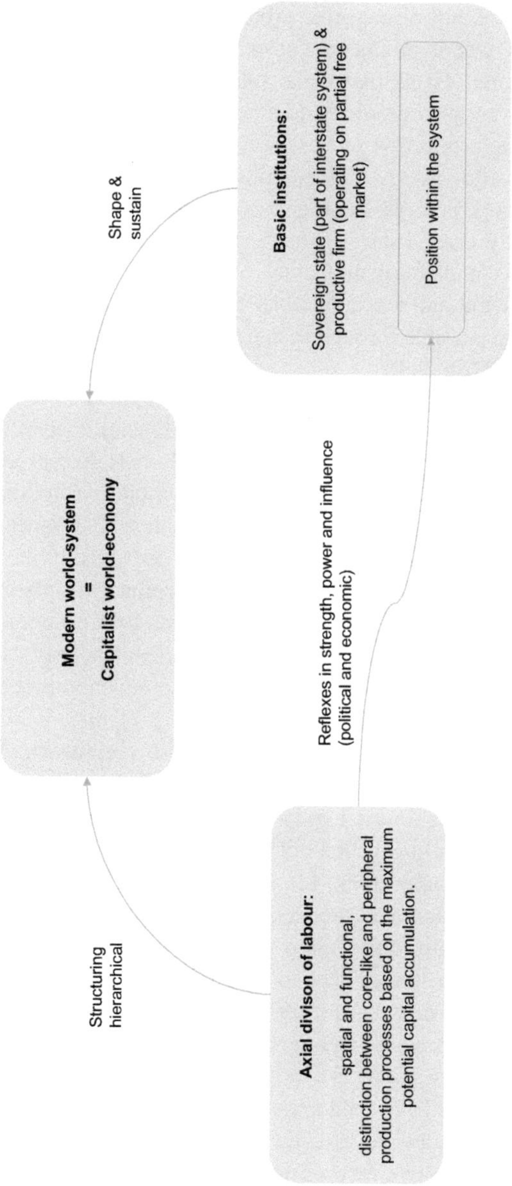

Fig 2.1 Visualisation of the components of the world-systems analysis relevant to this book

consumption and have little value (Wallerstein 2011a [1976], 302; 2011b [1980], 129). When the profitability of core-like production processes declines, they are relocated to regions where production costs are lower, thus becoming semi-peripheral or peripheral. Firms that make this shift transfer their accumulated capital from the now peripheral process to a new core-typical production. This is because only core-like production processes are profitable and therefore of interest in the modern world-system. The clothing industry is an example of axial division of labour. Work processes that generate high profits, such as design and sales, are now usually carried out separately from the cultivation of raw materials, their processing and the manufacture of clothing. These latter processes are considered peripheral and less profitable. The constant relocation of production processes is only possible through the spatial expansion of the system itself (Wallerstein 2011a [1976], 349).

The axial division of labour leads to unequal exchange, which was first described by Prebisch (1963) and assigned to dependency theory (Wallerstein 2006, 17). In this case, countries with predominantly core-like production processes benefit from the fact that products from the periphery (produced at lower production costs) are exchanged with those from the core (higher production costs) (Wallerstein 2006, 98). The monopoly position of core-like production processes allows for higher capital accumulation in regions with these processes. This results in the capability of states to generate power, both economic and political, within the interstate structure. According to Wallerstein, they can dominate weaker states, meaning those with mainly peripheral production processes (2006, 29).

The modern state in the modern world-system is a sovereign state. Sovereignty is, according to Wallerstein, 'a concept that was invented in the modern world-system. Its prima facie meaning is totally autonomous state power. But modern states in fact exist within a larger circle of states, what we have come to call the interstate system' (Wallerstein 2006, 42). The unifying principle of states is to maintain this interstate system and the modern world-system as a whole (Wallerstein 2006, 56). The strength of sovereign states is related to their position in the interstate system and can be traced back to the nature of the production processes based in the state. Strong states, according to Wallerstein, have a 'disproportionately share of core-like processes' and 'tend to emphasize their role of protecting the quasi-monopolies of the core-like processes', whereas states with a weak interstate structure tend to have predominantly peripheral production processes. As a result, these weak states are 'usually unable to do very much to affect the axial division of labor. and in effect are largely forced to accept the lot that has been given them' (Wallerstein 2006, 29).

2.3 The Relationship Between State and Productive Firms

The relationship between productive firms and states is fundamental to the capitalist world-economy, and thus to the structure of the modern world-system. Both share the aim of infinite capital accumulation and mutually promote their position within the system. Strong states are able to protect firms and their interests from other firms and states. However, a weak state and its withdrawal can also be advantageous for firms. This is the case, for example, when non-compliance with laws or regulations—for example, on nature conservation or workers' rights—goes unpunished. As a result, there is no state neutrality towards the production of value by firms within the capitalist world-economy. Thus, the maximum accumulation of capital by a firm is always a sign and a means of access to state protectionism. The promotion of quasi-monopolies combined with protectionist measures leads to the possibility for the firms concerned to escape the influences of the free market. The only way to achieve maximum capital accumulation. Through firms with monopolised production processes, a state generates capital in the form of tax revenue and thereby improves its own position in the interstate system (Wallerstein 2006, 18).

> *'[T]he model presumes that states regularly and unceasingly intrude their power and their ukases on the operations of the worldeconomy, and that they do so for a multitude of reasons: to maximize their own income; to respond to pressures of powerful entrepreneurial interests; to maintain political stability and minimize social revolt; to ensure their long-run military power' (Wallerstein 2000b, 197).*

Overall, the interplay between states and productive firms is fundamental to the structure of the modern world-system because '[t]he relationship of states to firms is a key to understanding the functioning of the capitalist world-economy' (I. Wallerstein 2006, 46). Although firm representatives often speak of their desire for a free market, they benefit greatly from protectionist measures when these are taken in their interest and thus give them the opportunity to develop into a quasi-monopoly. State measures that a state can take to promote the emergence of quasi-monopolies are, besides others, import and export restrictions on competing firms and their products, subsidies to the domestic firm or its ventures, or the purchase of products from the subsidised firm at inflated prices (Wallerstein 2006, 26). Another important instrument is the 'ability of strong states to use their muscle to prevent weaker states from creating counter-protectionist measures' (Wallerstein 2006, 26). These protectionist measures on the part of the state serve the purpose of helping the domestic firms to accumulate capital. External power is the instrument of state sovereignty that enables them, even at the request of companies, to exert influence and, if necessary, put pressure on another state to act in the interests of the firm (Wallerstein 2006, 50). The ability to control one's own national borders and thus set the rules for the movement of goods, capital and people is also an important instrument (Wallerstein 2006, 46).

Productive firms have a strong interest in states recognising their property and protecting it from access by others, so that they can continue to accumulate capital. Determining which costs a firm can externalise and which must be internalised is another crucial feature of state sovereignty. They decide which costs incurred or caused by the production process must be borne by the firm itself and are therefore part of its production costs and which are passed on to the population of the country (Wallerstein 2006, 47–48). Thus, capitalism is 'based on the constant absorption of economic loss by political entities, while the economic gain is distributed to "private" hands' (Wallerstein 2011a [1976], 348). In turn, Wallerstein describes property rights as 'the centerpiece of the capitalist system' because '[t]here is no way to accumulate capital endlessly unless one can hold on to the capital that one has accumulate' (Wallerstein 2006, 47). According to Wallerstein, the maximum accumulation of capital is 'a reward not merely for efficiency but for greater access to the assistance of the state' (Wallerstein 2006, 48). Because high profits allow firms to influence legislation and thus generate further added value. Wallerstein states that there is 'there exist no neutral positions for the state in enabling capital accumulation. For capital accumulation is always capital accumulation by particular persons, firms, or entities' (Wallerstein 2006, 56). It is therefore political decisions which firm promoted by state protection and gain more capital.

From the perspective of productive firms in the capitalist world-economy, the state is responsible for ensuring that firms can gain and maintain a competitive advantage. It achieves this by restricting or expanding the 'freedom' of the market. Whether the state actively intervenes or withdraws is irrelevant (Wallerstein 2011b [1980], 113).

2.4 Capturing the Structure of the World-Economy

Hopkins and Wallerstein supplement the research-oriented world-systems analysis with the concept of the commodity chain. The aim is to make the axial division of labour visible and analytically tangible by capturing and analysing these chains (Hopkins and Wallerstein 1977).

'We have wished to direct attention to what we consider to be a vital research location in the understanding of the functioning of the capitalist world-economy- the commodity chain' (Wallerstein 2000b, 199).

Economic structures can be described in various ways. For example, as economic fields (Bourdieu 1983; Fligstein 2001), organisational networks (Windeler 2001), global production networks (Coe, Dicken, and Hess 2008) or, as here, as commodity chains. Over the last decades, a large number of empirical and theoretical studies have been conducted on commodity chains. This research was accompanied by the development of the concepts of commodity chains (Hopkins and Wallerstein 1977) and global value chains (GVC) (Gereffi, Humphrey, and Sturgeon 2005). These two approaches differ in their research interest. However, both, the commodity chains approach and the global value chains approach, aim to shed light on today's (globalised) economy. The latter does this by focusing on the current situation, while the former focuses on the historical perspective. The aim of commodity chains concept is to provide an explanation of long-term, far-reaching social change (Wallerstein 2006). The research programme is therefore interested in a longue durée perspective. Summarising, while the global value chain approach focuses 'narrowly on the issue of *firm-level competitiveness* within the context of a particular industry, [it] sheds a very partial light on the critical question of winners and losers in today's global economy' (Bair 2005, 154). This results in the loss of perspective regarding larger (historical) contexts (Espeter 2025). Whereby 'world-systems theorists are most fundamentally interested in how commodity chains structure and reproduce a stratified and hierarchical world-system' (Bair 2005, 156). Therefore, in relation to my research question, the potential to explain global economic structures and their processes is high when empirical work from commodity chain research is conducted within the world-systems analytical framework.

A *commodity chain* is described by Hopkins & Wallerstein (1986, 159) as 'network of labour and production processes whose end result is a finished commodity'. They link production processes that are geographically dispersed: connecting regions, places and communities. Reconstructing a commodity chain requires the researcher to 'take an ultimate consumable item and trace back the set of inputs that culminated in this item—the prior transformations, the raw materials, the transportation mechanisms, the labour input into each of the material processes, the food inputs into the labour' (Hopkins and Wallerstein 1977, 128).

Examining commodity chains seeks to reveal the spatial extension of the axial division of labour in reality (Wallerstein 2000a, 11). Regarding the relationship between states and productive firms in the world-economy, it should be noted that commodity chains are 'trans state chains'. In other words, along the chain, commodities cross territories and national borders. This allows states to influence parts of the chains and the processes that take place within them (Wallerstein 2000a, 10).

World-systems analysis makes it possible to examine the process-nature of an economic structure and thus to draw conclusions about its continuity and change. Continuity within the modern world-system is achieved through the spatial and functional division of labour and its maintenance by states within the interstate system. Despite this stability, the system is cyclical. Consequently, Hopkins and Wallerstein's commodity chain framework can be used to capture both the process-nature of economic structures in the modern world-system.

2.5 Methodological Considerations for Commodity Chain Research

In addition to providing a theoretical definition of a commodity chain in the modern world-system, Hopkins and Wallerstein (1977) develop a research design for the empirical analysis of these chains. Their aim is 'to examine the geographic scope and social complexity of the commodity chains' (Hopkins and Wallerstein 1994a, 20). To this end, commodity chains should be studied at specific points in time. These are chosen on the basis of the A- and B-phases of the Kondratiev cycles (Kondratiev 1935) of the European economy during their study period between 1590 and 1790 (Hopkins and Wallerstein 1994a, 20). An A-phase describes expansion, while a B-phase describes contraction. First, the commodity chain and its process-nature are reconstructed. In addition, it will be examined 'whether and to what extent the structures of the boxes [authors note: production processes] change in accord with cyclical rhythms [A- and B-phases] of the world-economy' (Hopkins and Wallerstein 1994a, 20). In this research the focus is on the entire commodity chain (Wallerstein 2000b, 198).

By conducting commodity chain research within world-systems analysis, the empirical work offers a high degree of explanatory potential with regard to global economic structures and their process-nature. Hopkins and Wallerstein write: 'By tracing the networks of these commodity chains, one can track the ongoing division and integration of labor processes and thus monitor the constant development and transformation of the world-economy's production system' (Hopkins and Wallerstein 1994a, 17).

Hopkins and Wallerstein outline several research projects dealing with commodity chains. For example, they examine the spatial distribution of the commodity chain or of individual production steps, so-called 'boxes' (Hopkins and Wallerstein 1994a, 18). Following the concept of the axial division of labour, a distinction is made between core-like boxes 'located in a very few countries' und peripheral boxes which 'tend to have units in a large number of countries' (Hopkins and Wallerstein 1994a, 18). Furthermore, production units are relocated to peripheral regions as profitability declines (Hopkins and Wallerstein 1994a, 18). As described above, core-like production processes are highly monopolised. With regard to the formation of monopolies in commodity chains, Wallerstein states: 'the (relative) monopolies that are constructed are constantly under attack, and the more profitable the box the more steady the attack' (Wallerstein 2000a, 8). Consequently, the shifting of production processes and the resulting axial division of labour are fundamental to commodity chains (Hopkins and Wallerstein 1994a, 19).

In addition to studying the basic structure of a commodity chain and its spatial extension, Hopkins and Wallerstein advocate studying it in the longue durée: 'Furthermore, we shall not merely investigate the chains but we intend to investigate them at successive points in time, in order to see what kinds of changes took place in the construction of the chains, and to see if we can explain these shifts' (Wallerstein 2000a, 5). However, Wallerstein describes the study of commodity chains in the longue durée as a difficult undertaking, partly due to the availability of data: 'The kind of data we need is very hard to come by even for the contemporary period, and this is certainly so for earlier centuries. But the task is not hopeless, only very difficult' (Wallerstein 2000b, 198).

Since neither Wallerstein nor Hopkins provides any concrete solutions to this problem. I will demonstrate my methodological approach in order to close this gap in Chap. 3.

2.6 State of Commodity Chain Research Within the World-Systems Analysis

After conducting an extensive literature review, I state that, to the best of my knowledge, only two research projects have applied the theoretical and methodological concept of commodity chains introduced by Hopkins and Wallerstein. These projects therefore examine the process-nature of commodity chains in the longue durée. In addition to commodity chain research within the world-systems analysis, a number of other schools of thought exist. These were briefly highlighted in the introduction and will not be addressed further.

Focusing on commodity chains in the longue durée, the only studies were those of the Research Working Group 'Cyclical Rhythms and Secular Trends', coordinated by Hopkins and Wallerstein (Research Working Group on Cyclical Rhythms and Secular Trends 1979). Two contrasting cases were selected: the commodity chains of shipbuilding (Özveren 1994; Özveren 2000) and grain flour (Pelizzon 1994; 2000) (high-end or cheap everyday product). Furthermore, these studies adopt different approaches to analysing the investigation period proposed by Hopkins and Wallerstein: Özveren (1994; 2000) describes shipbuilding within periods, while Pelizzon (1994; 2000) considers the processes of the grain commodity chain along all periods.

Özveren (1994; 2000) reconstructs the commodity chain of European shipbuilding between 1590 and 1790. He does this by analysing the production steps necessary to build a ship, including inputs, but excluding labour, and neighbouring systems. In his process analysis, he examines developments over time in relation to the Kondratiev cycle. He compares shipbuilding sites along the European Coastline (f.e. in Arsenal, Venice, or Barcelona) and examines how internal (e.g., shifting production towards smaller ships) and external developments, such as access to resources such as construction timber, influenced each other. Focusing on the 'changing patterns of the chain', Özveren examines both the internal and external processes that result in changes to the commodity chain. The purely empirical article (1994) was later complemented by method and theory in a comprehensive article (2000). However, the connections to the extension and contradiction phases seem a little too short-sighted.

A significantly more complex reconstruction of the commodity chain can be seen in the case of grains. In her mapping, Pelizzon (1994; 2000) incorporates all stages of production, as well as inputs such as tools, labour, and land. She also considers 'alternative directions of the chain', such as storing the commodity along the chain instead of selling it, or selling products to an alternative market, for example for brewing instead of milling, and 'changing hands' between some production steps along the chain (Pelizzon 1994, 35). Initially, her description of the chain focuses on its various components (production, marketing, storage, consumption, etc.). Pelizzon maps the spatial structure of the commodity chain in each phase by: 'trac[ing] the geographic movements of grain in and out of major European cities in the seventeenth and eighteenth centuries' (Pelizzon 2000, 154). Like Özveren, Pelizzon examine processes based on the connections to the extension and contradiction phases.

In line with the theoretical concept of Hopkins and Wallerstein, Pelizzon focuses, like Özveren, on the 'changing patterns of the chain' (Pelizzon 1994, 39ff). She observes a 'tendency for the chains to lengthen geographically in B-phases' (Pelizzon 1994, 44). The 1994 article is merely a description and analysis of the empirical evidence, which is discussed in more detail in her 2000 article.

Both, Pelizzon and Özveren, considers 'trends' that initially have nothing to do with the economic fluctuations in the extension and contradiction phases, such as consumer behaviour—'an increasing tendency for white (wheat) bread to be consumed' (Pelizzon 1994, 34) or the spatial spread of alternative foods such as maize (Pelizzon 2000, 135). Nevertheless, both authors consider the combination of political influence on commodity chains, e.g. with regard to the availability of bread to prevent social unrest (Pelizzo 2000) or protectionist measures to support the domestic economy (Özveren 2000). Monopolisation and demonopolisation are highlighted as developments in expansion and contraction phases (Hopkins and Wallerstein 1994b).

In conclusion, I summarise that very little research has been conducted to date that applies the theoretical and methodological concept of commodity chains developed by Hopkins and Wallerstein to analyse the process-nature of the capitalist world-economy. One possible reason for the lack of research applying the concept developed by Hopkins and Wallerstein is that this kind of research is very extensive. Given the limited time and funding available for research projects, undertaking such a task seems challenging. In Chap. 3, I will first approach this research gap methodologically and present a concept influenced by Hopkins and Wallerstein. This concept will I then apply in Chaps. 4 to 6 to address the question of the process-nature of economic structures, using the cut flower commodity chain in the Aalsmeer region as a case study. Showing that this kind of research is possible and beneficial.

3 Process-Oriented Commodity Chain Mapping

> 'The processes of which we speak do not merely constitute the world-economy; they constantly reconstitute it' (Hopkins and Wallerstein 1977, 118).

In the following chapter[1], I will outline my methodological concept underlying the collection and analysis of economic structures within the modern world-system in order to obtain a comprehensive view of these structures.

To do this, it is necessary to take a close look at these economic structures, such as commodity chains, and their inherent power structures. However, in order to analytically process them, it is necessary not only to reconstruct the structure, but also to reduce its complexity. As I have already indicated (Espeter 2025), complexity reduction makes it possible to analyse structures as complex as those within the world-economy. For example, this makes it possible to identify inherent power structures. This is necessary in order to analyse the current state of an economic structure within the capitalist world-economy and its process-nature. However, to determine how the power constellations within economic structures have been formed, changed and maintained, it is necessary to go beyond the commodity chain mapping I introduced (2025) and adopt a process-oriented approach. In order to explore the emergence, continuity and change of economic structures, the methodological concept is carried out in the longue durée and thus in the process-nature of this structure: 'Without taking the historicity of social processes into account, researchers can neither theoretically conceptualise nor empirically

[1] Parts of this and the following chapters have already been published in Espeter (2025) in this or a modified form.

Supplementary Information The online version contains supplementary material available at https://doi.org/10.1007/978-3-658-52437-1_3.

L. M. Espeter, *From Hobby to Billion-Dollar Industry*, Wirtschaft + Gesellschaft,
https://doi.org/10.1007/978-3-658-52437-1_3

examine other social dimensions, such as power, knowledge, or space, as well as their entanglement' (Hergesell, Baur, and Braunisch 2020, 266).

In the following, I introduce process-oriented commodity chain mapping. I proceed as follows: (1) Selection of the reach design case study; (2) Case selection in process-oriented sampling taking into account product properties, space and time; (3) Data collection using mixed methods; (4) Data analysis involving the periodisation of the investigation period, the reconstruction of the structure of actors, the spatial structure and the inherent power structure within meaningful periods, and the subsequent analysis of the process-nature of economic structures within and along those periods. In the subsequent Chap. 4, I provide an explicit explanation of the aforementioned concept, with the case study of the cut flower commodity chain being used to demonstrate the methodological approach.

3.1 Research Design: Case Study

Although the research process is usually carried out iteratively, in the case of commodity chain research in the longue durée, it is necessary to carry it out linearly due to the high level of complexity (Baur 2019). The objective of selecting a research design is to address and answer the underlying research question. This prompts the following question: How can research questions on economic structures and their process-nature be implemented in empirical social research?

A process-oriented approach to commodity chain mapping (Espeter 2025) emerges as a necessity when considering process-related aspects within the context of economic structures. To this end, I introduce the investigation of process-related aspects of economic structures as process-oriented (Hergesell, Baur, and Braunisch 2020; Baur 2017) case study research (Yin 2018).

Case study research is a comprehensive research strategy. In this strategy, it is essential to follow the many facets of the case and to present it in its complexity as comprehensively as possible. Case studies do not focus on a specific data collection or analysis procedure. Rather, the focus is on a comprehensive research design with the goal of achieving an in-depth understanding of the case (Hering and Jungmann 2022). A case can be not only an individual or an organisation, but also a social process (Baur and Lamneck 2017). The distinction between the case and its environment is fluid (Hering and Jungmann 2022). Therefore, it is crucial to pay particular attention to case selection.

3.2 Case Selection: Process-Oriented Sampling

Every research project begins with the definition of the case. As the name suggests, a case study is on a single case. With regard to the consideration of economic structures and their process-nature, the case can be selected, for example, on the basis of a product, a specific commodity chain, or a process. For example, the product 'cut flower', the Kenyan cut flower commodity chain from Lake Naivasha to Berlin, Germany, or the process of producing cut flowers, including all inputs. The selection is made intentionally (Baur and Lamneck 2017; Yin 2018). The case is selected on the basis of its content, in order to be able to answer the research question.

It is important to emphasise that the selection of cases affects the generalizability of the research findings (Akremi 2022). The purposeful selection of cases is either a typical case (average, representative), an extreme case (stands out), or a verification case (in contrast to other cases already studied). For example, Dutch flowers for the German market (typical case), banana production in Iceland (extreme case), or contrasting production systems e.g. conventional/organic or traditional/high-tech (verification case).

In quantitative social research, the size of the random sample in relation to the population is considered an indicator of how generalisable the research is. In qualitative social research, however, the principle of theoretical saturation applies. This involves selecting cases iteratively before and during the research process until no new insights into the underlying question can be gained. Once theoretical saturation has been achieved, generalisations can be made beyond the cases.

In terms of thematic sampling the data units (*Erhebungseinheiten*) (Akremi 2022), three levels are relevant for commodity chain mapping: the product, the location, and the time. I will outline these in the following section.

3.2.1 Product Properties

The first step is to define the case, in this case the commodity chain through the product. The research question to be answered serves as the basis for this decision. In order to analyse the process-nature of economic structures, such as commodity chains, it is necessary to select the case according to their product characteristics and process steps. In order to gain an in-depth understanding of the chain and its structural characteristics given by the product itself from the outset, it is necessary to consider the product properties and process steps of the commodity chain. In this step of case selection, the differences between the commodity chains of different products are revealed. This is because for example the commodity chains of coffee (Talbot 2011), ships (Özveren 2000), phones (Lee, Gereffi, and Lee 2016) or clothing (Gereffi, Spener, and Bair 2002) vary greatly. These differences are due, for example, to the number and type of production steps required. Is it an industrial or agricultural product? Are separate processing steps required? In addition, product properties such as weight, perishability or fragility also influence the requirements for the steps along the commodity chain. The target markets of the product also influence the commodity chain (Dobeson and Kohl 2020). Is the product traded in a primary market (e.g. fresh produce), is there a secondary market (e.g. cars, jewellery, art) or is it part of a circular economy (e.g. waste, clothing)?

3.2.2 Space

Once a product has been selected and familiarised with, the next step is to determine the spatial focus of the study. This involves deciding which elements of the chain to examine and where to conduct the study.

With regard to the elements of the chain, it must be determined whether the focus is on the commodity chain as a whole, including all production steps and inputs, or on individual parts of the chain or if the focus of the study, and thus the case selection, should be on individual parts, such as input, production, logistics or consumption.

The principles of conscious case selection are also applied here: Should a typical, a standing out or verification location be examined? To elaborate further, when selecting a rose commodity chain based on its production location, should the starting point be typical glasshouse production, as in Kenya (a typical case), or open-field cultivation of cut flowers, as in Germany (an extreme case), or Colombian flower production as opposed to Kenyan production (a verification case)?

3.2.3 Time

The challenge in considering case studies in the longue durée is that the case changes over time (Norkus and Baur 2020, 3–4; Hergesell, Baur, and Braunisch 2020, 272). This may relate to the product itself, the steps necessary for its manufacture, transport or distribution, or the physical spaces it passes through in the commodity chain. It is therefore necessary to derive the temporal frame of reference for considering both the case itself and the underlying research question: 'The actual sampling procedure can begin after researchers have considered duration and temporal patterns, decided on a time frame and concluded periodization' (Hergesell, Baur, and Braunisch 2020, 280).

Consequently, the temporal boundaries of the case must be defined in the analytical step of periodisation. In order to observe process-nature or 'temporal patterns' (Baur 2005, 125–63), the time under study should ideally begin 'before or when the social process of interest begins; it ends when or after the social process ends' (Hergesell, Baur, and Braunisch 2020). However, '[t]rajectories often do not have a fixed beginning and ending' (Norkus and Baur 2020, 4). Therefore, according to Baur (2017), it is necessary to carry out the steps of backward and forward reading for historical reconstruction. Backward reading, 'traces the figuration back in time [....] to the formative phase' (Baur 2017, 54).

In the case of economic structures in the modern world-system, this formative phase can be characterised by a variety of things. For example, the focus was on the commodity before it became part of the commodity chain. In the case of cut flowers, this would be when flowering plants were used for medicinal purposes or as decorative elements in gardens. Or, if the focus is on a particular location of production, the location before it became a location of production, whether through the establishment of a firm, the formation of a cluster, or a relocation. Or the situation before, for example, an external shock (Hering and Espeter 2025) influenced the initial economic situation in one way or another.

The time under study is then divided into periods. Baur (2017) refers to periodisation as a separate analysis step, the more detailed description of the implementation follows in Chap. 4.

At the sampling level, however, I like to note that there are different ways to implement periods. One commonly used method is to divide the time under study into periods of equal length, such as decades, years, or months. Hopkins and Wallerstein (1977), on the other hand, base their research design on the Kondratiev cycle (Kondratiev 1935). The Kondratiev cycle describes developments in terms of two phases. A-Phases describe expansion, while B-phases describe contraction. The duration of the cycles 'depends on the political measures taken by states to avert a B-phase' (Wallerstein 2006, 31). The division into periods is therefore based on periods of contraction and expansion (Hopkins and Wallerstein 1977, 167). A third option is periodisation based on *meaningful periods* (Baur 2017, 54), which I consider to be particularly valuable for the purposes of this kind of research. These periods are chosen so that the structure under consideration is relatively stable within the period. Forward reading is used for this classification. Starting from the formative phase, forward reading identifies structural changes and thus content-based periods (Baur 2017).

3.3 Data Collection: Mixed Methods

Subsequent to the collection of data, the relevant data sources must be identified. In terms of interviewees, it is important to avoid using convenience sampling, such as snowball sampling (Akremi 2022; Baur and Christmann 2024).

When collecting data, a conscious decision must be made as to which data sources are appropriate within the theoretical framework to answer the research question. In considering the longue durée, it is necessary to take into account a wide variety of data. For example, the collection and use of one's own interview data or other research-elicited data (Graeff and Baur 2020) is a reliable source of data only as far back as human memory can go. Further back, process-generated data are needed (Hergesell, Baur, and Braunisch 2020). Data availability is a major challenge, as often not all types of data are available in all periods (Norkus and Baur 2020). Therefore 'in process-oriented research, mixed methods [...] often are not a choice but become a necessity in order to be able to address the research question' (Norkus and Baur 2020, 12). It is therefore necessary to consider the use of different forms of data. The data mix can be purely qualitative or quantitative. However, a qualitative/quantitative mix is often useful. Depending on the time under study, the forms and sources of data will differ from period to period. In terms of the individual steps in the data analysis process, compromises may have to be made and certain aspects of the analysis may have to be neglected due to the availability of comparable data.

3.4 Data Analysis: Reconstructing Economic Structures and Processes

In order to observe the inherent processes of economic structures in the modern world-system, the first step in data analysis is to divide the period under study into meaningful periods. Further data analysis is carried out after the periodisation within these meaningful periods.

Therefore, after the backward reading and identification of the formative phase, the first step of analysis is to define the periods. The division of the time under study into periods is done to reduce the complexity of the historical analysis and the necessary data material. In addition, the investigation is sharpened by the so-called periodisation (Hergesell, Baur, and Braunisch 2020). The subdivision of the time under study can be done, among other things, into equal periods (years, decades, etc.), along economic cycles (Hopkins and Wallerstein 1977) or meaningful periods (Baur 2017).

Influenced by my empirical results during this study and by the work of Hergesell (Hergesell 2019; 2021), I advocate a periodisation based on content. These meaningful periods are formed and examined during the historical reconstruction. In this context, they are not rigidly limited in time. In other words, processes that establish these periods sometimes run parallel to each other. Looking at processes beyond macroeconomic fluctuations (see f.e. Kondratiev 1935) makes it possible to explicitly consider their effects. The periods are deliberately set on the basis of the underlying research question and iteratively through the analysed data itself.

With regard to the process-nature of economic structures—such as commodity chains—in the modern world-system, the same questions are asked of the data material in each period. The central processes taking place in one period are characteristic to that period. To this end, events of emergence and change are considered.

The analytical focus is on the process-nature within and across periods. In order to analyse the structure of the commodity chain under study, the data is examined in three successive steps for each period: structure of actors, spatial structure and inherent power structure.

Depending on the research question and case, the weighting and scope of the respective analysis steps are to be set differently. For example, if the focus is on a specific production site, the spatial extension of the commodity chain plays a subordinate role. If the focus is on a specific actor, it may be sufficient to consider only a reduced actor structure in each period, based on a complete reconstruction, as I have analysed elsewhere (2025).

3.4.1 The Structure of Actors of Commodity Chains

The periodisation is followed by the reconstruction of the structure of actors. As with the commodity chain approach (Hopkins and Wallerstein 1986) and global value chain mapping (Frederick 2021), this reconstruction includes all the necessary steps from the input material to the finished product in the final consumer market. The focus of commodity chain mapping (Espeter 2025) is on the actors involved. The reconstruction follows the scheme of a reduced commodity chain illustrated in Fig. 3.1. It integrates input materials and processes, production processes, transport and logistics, world market and consumer market. Labour-power and its reproduction is the foundation of the entire system (Hering and Espeter 2025). Education, environment, politics etc. are neighbouring systems to the chain (Baur 2023).

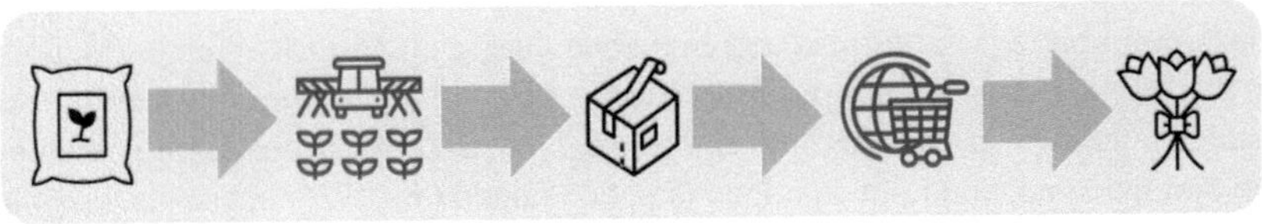

Fig. 3.1 Reduced structure of actors of an agricultural commodity chain (own illustration).

This analysis step requires a reduced reconstruction. To continue working analytically, reducing the structures complexity is necessary. Thereby, it is necessary to identify the fundamental components of the chain that are essential for the entire chain and all its steps. Without these actors, the chain as such could not exist and therefore no accumulation of capital would take place. These include typically labour and its reproduction, supported by social infrastructure like schools, health care, local communities etc. and natural resources which are extracted from the nature (e.g. water, raw materials) and built infrastructure (roads, energy etc.) which were often initially established and funded by state actors. An example of how this reduction can be implemented can be found in Espeter (2025).

3.4.2 The Spatial Structure of Commodity Chains

In the application of world-systems analysis, the unit of analysis is the modern world-system. The spatial demarcation between cases and their comparisons is thus not based on political, i.e. nation-state, borders (Wallerstein 2006). Rather, it is based on content. With regard to examine the process-nature of the world-economy, Hopkins and Wallerstein (1977) among others, formulate the research task of 'Bounding the World-Ecomomy'. The goal is 'mapping the geographical extent of the world-economy *over time*' (Hopkins and Wallerstein 1977, 122—emphasis added by the author). To do this, it is necessary to define methodologically and empirically which parts are inside and outside the world-economy at which point in time and thus in the period under study (Hopkins and

Wallerstein 1977). It is precisely this that must be applied to the study of the process-nature of economic structures.

Therefore, in the next step the spatial structure of the commodity chain is analysed. For this purpose, all actors are analysed in terms of their position in space. This begins with the geographical position and leads to the position in the modern world-system. This involves considering, on the one hand, where the actor is located in geographical space and, on the other hand, what is located around them. Which economic environment is it in, is it a region with mainly core-like or peripheral production processes? Which state and supranational actors could potentially influence it? These factors have a major impact on the inherent power structure of the commodity chain.

3.4.3 The Inherent Power Structure of Commodity Chains

The world-systems approach inherently include power. However, this approach does not formulate, or only inadequately defines power. It is therefore essential to develop a coherent concept of power in order to consider power in commodity chains (Espeter 2025).

Power relations in commodity chains can be analysed on at least four levels: between individual chain links (Gereffi, Humphrey, and Sturgeon 2005; Dallas, Ponte, and Sturgeon 2019; Ponte, Sturgeon, and Dallas 2021), between competing production sites (Massey and Meegan 1982; Massey 1995), between competing production systems and markets (e.g. primary and secondary markets, made-to-order and made-to-stock, organic and non-organic) and along the entire chain. For these relationships, it is necessary to integrate a coherent concept of power into the analysis based on the research question.

3.4.4 Process-Nature of Commodiy Chains

Following the reconstruction and analysis of the structure of actors, spatial structure and inherent power structures of commodity chains within the modern world-system, it is necessary to examine the processes involved.

Following this, my methodological concept is focusing on the processes between and within the periods. It is first necessary to create an event matrix in which processes are assigned. According to Baur (2005), the creation of this event matrix an interpretative act, as the researcher must use the data to identify events and separate important from unimportant information (Baur 2005, 118).

The columns of in the event matrix (see appendix 2) are divided into the meaningful periods identified in the periodisation and thus represent the temporal sequence. These meaningful periods are not rigidly limited in time, and the processes that give rise to them sometimes run in parallel. Nevertheless, this categorisation makes it possible to distinguish between before, during and after. The rows indicate the thematic criteria for comparison based on the underlying research question.

4 Methodological Implementation: Cut Flowers as a Global Commodity

Having introduced process-oriented commodity chain mapping, I will now explain how I applied it to my study of the cut flower industry. To answer the underlying research questions about the process-nature of economic structures and the question of how power is established and maintained in these structures, I examine the cut flower commodity chain in a process-oriented case study in the longue durée. The cut flower sector with its spatial and organisational centre of the flower auction in Aalsmeer serves as the empirical case study.

The cut flower sector offers a variety of promising aspects when examining the connection between continuity and change, due to its long history and connection to the globalisation of agricultural production processes, among other things. Moreover, the evolution of this commodity chain is particularly remarkable in view of the fact that a central actor, the flower auction in Aalsmeer, has emerged. Furthermore, this case makes it possible to analyse the relocation of the production of a perishable product and its effects on the structure of the commodity chain in terms of actors, space and power. In doing so, I advocate for a historical and global perspective on economic relations.

Building on the detailed reconstruction and abstraction of the current structure of the commodity chain of Kenyan cut flowers (Espeter 2025), in this work I focus on the process-nature of this large economic structure. In doing so, an explanation for why the structure is 'becoming what it is and not something else' (Weber 2002 [1904]) is found in the historical examination of the central actor—the Dutch flower auction in Aalsmeer.

L. M. Espeter, *From Hobby to Billion-Dollar Industry*, Wirtschaft + Gesellschaft,
https://doi.org/10.1007/978-3-658-52437-1_4

I achieve this through the following steps: (1) selecting and describing cases; (2) selecting and collecting data. This is then followed by the following analyses: (1) periodisation (2) reconstruction of the structure of actors, spatial structure and inherent power structure within meaningful periods, and (3) analysis of the process-nature of commodity chains throughout the time under study. Figure 4.1 visualises the research process.

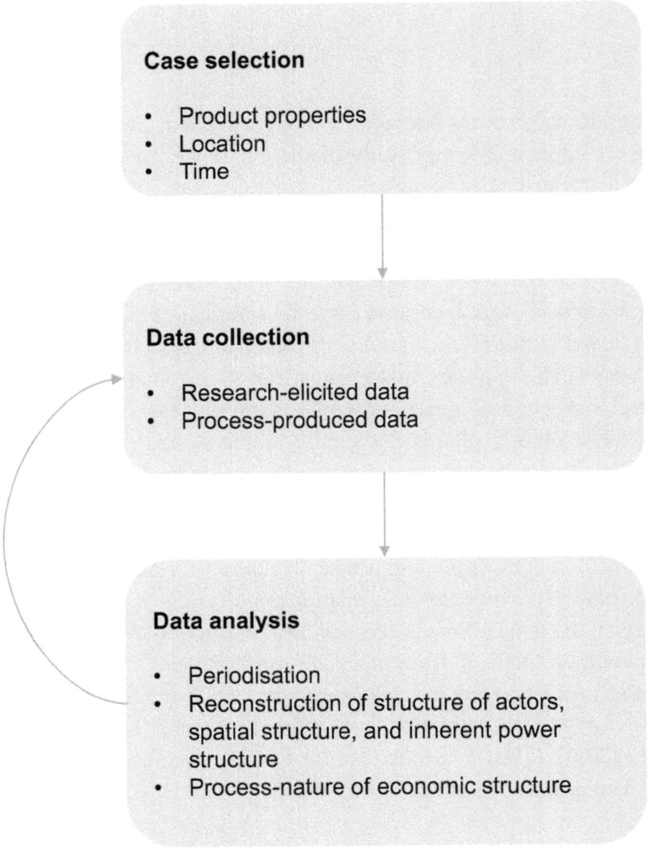

Fig. 4.1 Visualisation of the research process. (Own illustration)

I argue that changes to and stability of economic structures are accompanied by various internal and external processes. These include technological innovations; the restructuring of chains by new actors; external crises and shocks; and political influence. In addition to economic rhythms (Kondratiev 1935; Hopkins and Wallerstein 1977; 2016), external shocks (Hering and Espeter 2025) and, above all, processes of establishing and maintaining power influence the process-nature and therefore the economic structures.

Consequently, the selected case study is used to examine the world-economy within the context of the modern world-system. It also illustrates the axial division of labour by examining the inherent structure of the world-economy in the form of commodity chains. Furthermore, the longue durée approach allows us to analyse the process-nature of these chains.

4.1 The Case: Cut Flower Commodity Chain

Before beginning the empirical analysis, it is necessary to lay out the case selection, beginning with the product properties, following by the location and last time.

4.1.1 Product Properties of Cut Flowers

The first question that arises is what is the product in the cut flower commodity chain? Is it a cut flower or, for example, the bouquet/arrangement made from it? Wallerstein has already noted the 'ambiguity of the term "finished product"' (Wallerstein 2000a, 3). Moreover, a flower cut from a plant is not immediately a cut flower. It has to be commodified first in a historical process, which I will discuss in Sect. 5.1 in the examination of the formative phase 'Imperialism and Curiosity (1570s to 1850s)'. The cut flower industry is characterised by the materiality and rapid perishability in the fresh produce sector. Proper handling and refrigeration are critical to the shelf life and quality of the product. Furthermore, unlike industrial products or fast-growing plants, cut flowers cannot be easily reproduced. Demand must be anticipated and calculated well in advance in the production process. Before for example rose bushes can be grown on cut flower farms, varieties are developed over years and decades. When these varieties meet expectations and trends, cuttings are produced. These in turn grow for months before the first cut flower can be harvested. Rose bushes are productive for cut flower production for about three years. A detailed description of the production process of roses in: Hale and Opondo 2005; Riisgaard 2009; Nowakowska 2015; and Engel 2025. In addition, most cut flowers can only be stored and transported for a short time and under certain conditions. They are transported without water and must be stored in a cool place (around five degrees Celsius). Due to their fragility, the cut flowers also require secure packaging. Both require careful and fast logistics to ensure that the cut flowers reach the end consumer in the best possible condition. This means that the product itself influences production and logistics, while production is highly dependent on the market. Unlike many other agricultural products, cut flowers are not an everyday commodity, but a luxury item consumed mainly in regions of the core (e.g. Germany, US, Japan). However, as is often the case with agricultural production, the potential for value addition is low. Production is labour- and cost-intensive and dependent on a wide range of environmental factors. As my analysis will show, the position of actors in the structure of the commodity chain is crucial to their capability of capital accumulation.

The demand for cut flowers, and therefore their production, is driven by calendar peaks. These are mainly in February to May, around Valentine's Day (14 February), International Women's Day (8 March), Mother's Day (first Sunday in May) and British Mother's Day (fourth Sunday in Christian Lent). In addition to organising production for such annual sales peaks, production regions are closely linked to and therefore highly dependent on specific sales markets. This is partly due to geographical proximity, but also to specific market preferences. For example, around 80% if the cut flowers produced in Colombia (mainly roses (19.1%), carnations (13.4%) & chrysanthemum (8.6%) (Foral Daily 2023)) are sold in the US (OEC, n.d.). In contrast, the most popular cut flower in the US market, the tulip ('US: Best Selling Cut Flowers 2019', n.d.), is produced in the US itself. The necessary bulbs are imported as inputs from the Netherlands (Rabobank 2022). The German market is dominated by demand for roses (38% in 2022 (Schirmer 2024)). The main areas of production for the Germany market are the Netherlands and Kenya (Schirmer 2025). Before Russia's war of aggression in Ukraine, yellow roses were grown in Kenya that were popular in and destined for the Russian market. These varieties have been difficult to sell since the trade embargoes imposed in response to Russia's actions (Farm Manager 2022).

4.1.2 The Cut Flower Commodity Chain in Space and Time

The cut flower commodity chains are extensive. There are at least two ways of looking at the spatial distribution of cut flowers: (a) focus on a single commodity chain or (b) focus on flow of goods between production site and sales market. Although the two approaches are very different in terms of methodology, they both reveal the spatial and functional division of labour between core and periphery. Focusing on a single chain (a) maps the complexity commodity chains. Depending on the underlying question, the focus is on which work and production steps are necessary, who the central actors in this commodity chain are and, in this case, where the inputs for production come from. For example, I have shown (2025) that in cut flower production in Naivasha, Kenya (one of the largest rose production sites), inputs come from China, Russia, the Netherlands or even Israel. Research and development, on the other hand, takes place both in the Netherlands and locally. This axial division of labour is reflected in the unequal exchange between core and periphery.

Focusing on the flow of goods (b) illustrates the spatial dimension of the world market and the relationships between production and distribution. Process-based indicators, such as import and export data, are used as data sources.

Globally, there is a large market for cut flowers. In this context, wealth plays a major role in the sale of this consumer good. In 2022, 62.1% of the total market was imported by the top five importers: Germany with 13.5%, the Netherlands with 10.9% and the UK with 8.9%. The US is the world's largest importer, with a market share of 25.1%. In terms of the value of goods, the top five importers (including Russia, which accounts for 3.7%) on the world market as a whole import cut flowers worth about $6 billion. Local cultivation and distribution not included. In contrast, the countries of South-East Asia together import around 10%. Australia, the largest country in Oceania, imports 0.9% and Chile, the largest South American importer, imports 0.4% of the world market. Countries of the African continent each import less than 0.1% (International Trade Centre 2025b).

In contrast to consumers, producers are often located in regions that can be described as peripheral. In terms of the value of exports, the Netherlands was the world's largest exporter of cut flowers in 2022, with a trade value of USD 4.9

million[1]. This was followed by Colombia with USD 2 million, Ecuador with just under USD 1 million, Kenya with around USD 670 thousand and Ethiopia with around USD 230 thousand. Measured in terms of the quantity of goods exported (here in tonnes)[2], the picture is quite different. In 2023, the Netherlands exported 594 102 tonnes of cut flowers, Nigeria 539 440 tonnes, Colombia 306 623 tonnes, Ecuador 180 905 tonnes and Kenya 179 875 tonnes. The large difference between export values and export volumes can be quantified in terms of the value of the goods. For example, a tonne of cut flowers from the Netherlands was worth an average of USD 8280, while Nigerian cut flowers were worth USD 271 per tonne. This puts Nigeria at the bottom of the list, together with El Salvador (USD 222 per tonne) and Nicaragua (USD 193 per tonne). In terms of quantity, El Salvador (18 tonnes) and Nicaragua (955 tonnes) cannot compete with the world's largest exporters. Products exported from Italy (USD 8138/tonne) are comparable to the Dutch value of flowers (International Trade Centre 2025a). This illustrates that cut flowers from core regions have higher product values. However, it is not possible to determine in each case whether this also means that the generated surplus is also higher. As production costs are also higher in core regions.

Another remarkable aspect of the global flow of cut flowers is the distance between the producing country and the importing country. Focusing again at the five countries with the highest exports (by export value), the global arrangement of the different chains becomes clear. Cut flowers exported from the Netherlands are transported an average of 1,138 km to the importing country, Colombian flowers 5,231 km, Ecuadorian flowers 7,964 km, Kenyan flowers 6,357 km and Ethiopian flowers 5,296 km (International Trade Centre 2025a). The spatial differences between core and periphery in terms of production and consumption (Australia excluded) are becoming apparent. Figure 4.2 illustrates the global flow of cut flowers. It shows that there are clear destinations for the specific production regions.

For example, the flow of Colombian cut flowers is mainly to the US and Canada, while Kenyan cut flowers are destined mainly for the European market.

At the same time, in addition to commodity cycles, there are regional production areas where there is no great geographical distance between production and sales markets. On the one hand, these are European countries, as is the case for the European market. On the other hand, there is domestic production. These are

[1] As the Netherlands is an important re-exporter of cut flowers, the export figures also include re-exports. Domestic consumption would need to be taken into account. However, no reliable figures are available.

[2] The USA and Australia are excluded here, as the figures for these countries are only available in units and are therefore not comparable.

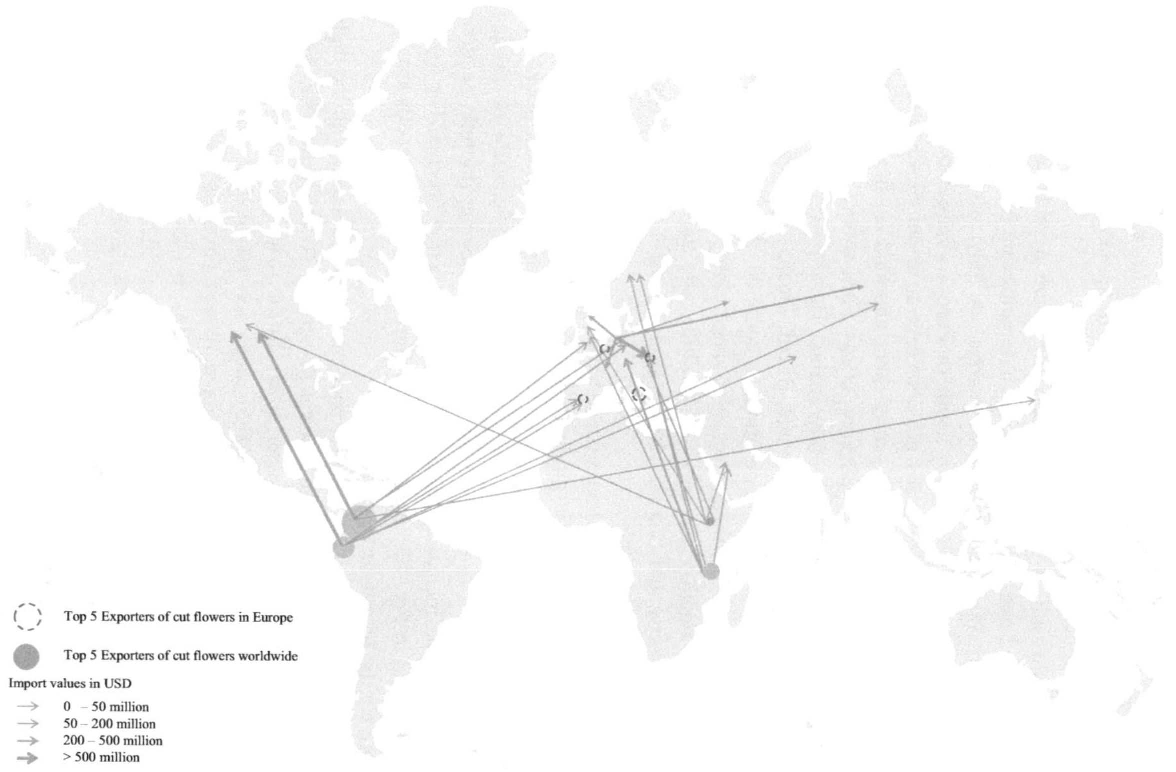

Fig. 4.2 Visualisation of the flow of cut flowers from the top five global importers and exporters. (Own illustration, based on: The Observatory of Economic Complexity. n.d.-a)

not included in the export/import statistics. The Indian, Chinese and Canadian cut flower commodity chains can therefore be used as counter-examples to the globalised cut flower industry. The area under cultivation in India is around 31,000 hectares and 15,000 hectares in China and therefore the largest production areas worldwide (Gemählich 2022, 52).

Since the beginning of the cut flower trade, there have been central transhipment points and marketplaces. These include wholesale markets, such as the 'Mercato dei Fiori Sanremo' (wholesale market) in Italy, which is the largest transhipment point for cut flowers in Southern Europe ('Sanremo Flower Market', n.d.). However, logistical hubs are also central to cut flower commodity chains. In the US, for example, two out of three imported cut flowers arrive via Miami airport and 'PortMiami' (Vellekoop 2024). Furthermore Flower Auctions are also central to cut flower commodity chains. These flower auctions exist in the Netherlands, Germany, China, Canada, Brazil, Japan, New Zealand and Taiwan (see Fig. 4.3). They have several things in common: they are usually owned and operated by the growers themselves, all located close to airports, surrounded by a cluster of production and logistics facilities, and the goods traded there are distributed directly and 'via the clock'. Many of them also have (personal) connections with the Netherlands.

The Dutch auction system, often referred to as selling 'via the clock', is a form of auction in which the price is gradually reduced from a high starting value, usually defined by the auctioneers, until a bidder accepts the current price. This type of auction is characterised by the use of a clock to indicate the decreasing price. Multiple clocks can be active simultaneously, allowing multiple auctions to run in parallel and increasing the efficiency of the auction process. The flower auctions (mapped in Fig. 4.3) are moreover trading centres where goods are sold directly or via the Dutch auction system.

In Canada, there are two main flower auctions in the provinces with the highest production: Ontario (223.51 million cut flowers) & British Columbia (69.92 million) (Statista Research Department 2024). In the main production area, the Ontario Flower Growers Auction in Mississauga was established by farmers in 1972 (Ontario Flower Growers. n.d.). United Flower Growers was founded in the 1970s in Vancouver. It sells mainly (90–95%) to local growers in British Columbia ('UFG Customer Access: About the Auction' 2025). The sales markets of this trading centre are Western North America ('Welcome to United Flower Growers' 2025).

The Chinese flower auction Dounan Flower Market (斗南花卉市场 Dǒunán Hāhuì Shìchǎng) is located in Kunming. This region has the largest flower production in China ('Kunming' 2025). The flower production was built up

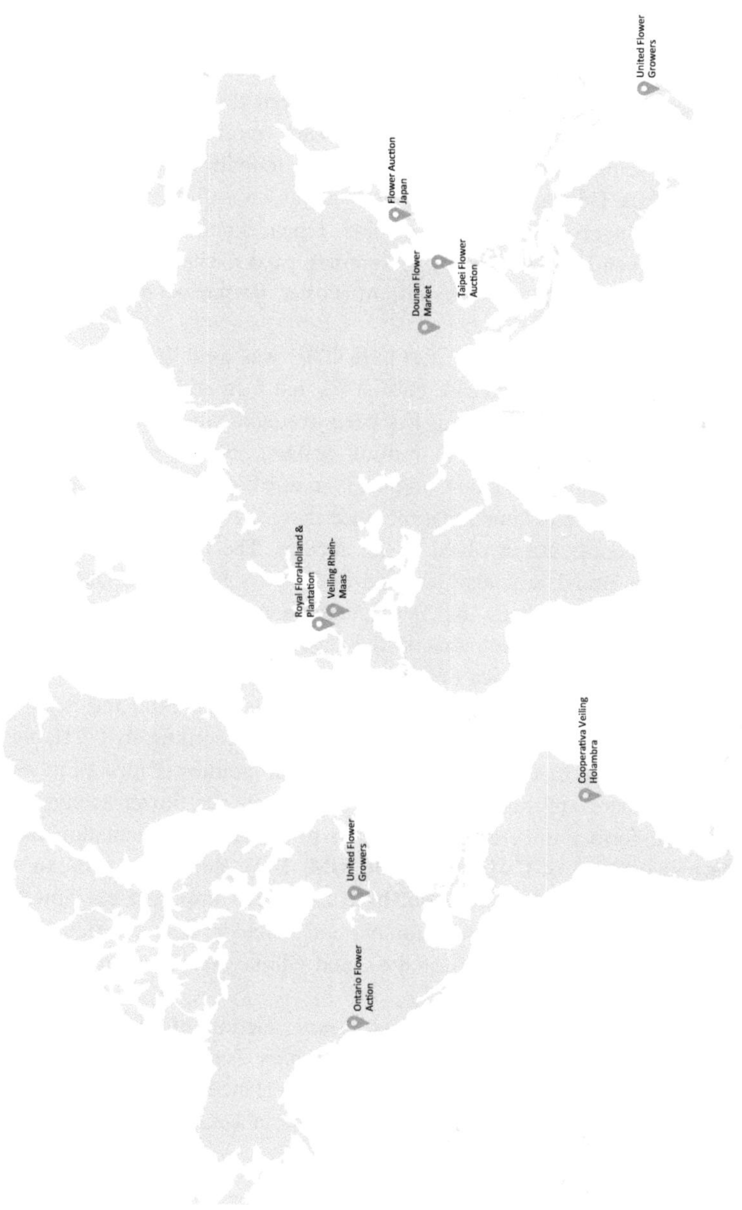

Fig. 4.3 Flower auctions worldwide. (Own illustration)

in cooperation with Dutch growers. The proximity to Dutch producers is also expressed through the town friendship since 2013 with Alkmaar in North Holland ('Kunming' 2025). At the Dounan Flower Market, a combination wholesale market and auction, 80% of China's cut flowers are traded (Rong 2022). Unlike the other auctions (which operate mainly in the early morning), this market is only open in the evening and is also open to private individuals (Li 2024). The annual turnover of the Dounan Flower Market is about 2.5 million renminbi yuan (about €325,000). Buyers come from the China, Japan, South Korea, Singapore and Thailand ('Dounan Flowers Market in Kunming', n.d.). This is also reflected in the regional orientation of the nearby airport (Porter, Ramirez-Vallejo, and van Eenennaam 2013, 15).

The Taipei Flowers Auction (台北花卉批發市場) was established in Taiwan in 1988 (Christian/Western calendar) ('市場沿革', n.d.). In addition to auctions under the Dutch system, online trading has been available since 2018. In 2022, Taiwan exported cut flowers worth 41 million dollars, 85% of which went to Japan (The Observatory of Economic Complexity. n.d.-b). As there is no English-language website, it can be assumed that the trade there is for the regional market and that no Dutch people are involved in the auction. The Cooperativa Veiling Holambra (CVH) in Brazil, was also founded in 1989 by farmers of Dutch origin (Ploeg 2019; Paprika Filmes, n.d.). It is also a cooperative ('O Veiling', n.d.) and is located in the region that is the largest producer of flowers and ornamental plants in Latin America (Wiersinga et al. 2013, 7). The municipality of Holambra was founded as a colony by Catholic Dutchmen after the Second World War (Oliveira, Pozzebon, and Heck 2007, 41). According to CVH, the cultivation of flowers can be traced back to three Dutch families (Paprika Filmes, n.d.). Until today there are strong links with the Dutch cut flower sector, for example with the former director of the Dutch flower auction Plantation taking over the management of CVH (Gebhardt 2014, 102). Brazil's import/export balance illustrates domestic consumption. In 2022, Brazil imported cut flowers worth USD 2.124.000 (mainly from Colombia and Ecuador) and exported USD 119,000, mainly to the Netherlands (58.8%), and China (32.8%) (International Trade Centre 2025b).

Furthermore, there is the New Zealand flower auction of United Flower Growers in Auckland, Christchurch, Wellington, in New Zealand (United Flower Growers, n.d.), and the Flower Auction Japan close to Haneda Airport. The latter is the second largest wholesale market for cut flowers in Japan ('Flower Auction Japan', n.d.).

In the Netherlands, there are two flower auction cooperatives: Plantation and Royal FloraHolland. Plantation in Ede is a 'smaller and more local' auction (Gebhardt 2014, 31). Royal FloraHolland is the largest flower auction in the world (Beemster 2017, 4; 'Largest Flower Auction', n.d.). Royal FloraHolland is operating auctions located in: Aalsmeer, Naaldwijk, Rijnsburg, Eelde and Veiling Rhein-Maas (Germany) (see Fig. 4.4). Royal FloraHolland also has offices in China, Kenya and Ethiopia (Royal FloraHolland, n.d.). The marketplace for international trade is located in Aalsmeer, Naaldwijk and Rijnsburg. In Eelde and the German Veiling Rhine-Maas, the focus is on the regional Dutch and German markets (Owner 2022b). Royal FloraHolland Aalsmeer is located close to Schiphol International Airport and Amsterdam (see Fig. 4.5).

In 2022 Royal FloraHolland had an annual revenue of 2022 5,166 Million Euro. The auctions were supplied by 4885 suppliers, of which 3415 were members of the cooperative. 43,442 thousand transactions were made by 2,333 buyers. Of which 40% via clock auction, 60% direct trade (Royal FloraHolland 2023). At the various Dutch locations of the Royal FloraHolland there are auctions that take place physically (Rijnsburg, Elde) or online (Aalsmeer, Naaldwijk). Physical means that both the goods and the potential buyers are in the auction room at the time of the auction. For direct trade, Royal FloraHolland is running the online platform Floriday. In 2021, Floriday was the largest digital B2B marketplace in Europe, with over 1,700 producers and direct sales totalling €1 billion (Royal Flora Holland 2022).

Despite the size and role of the cut flower market, the flower auction in Aalsmeer is quite modest. Its sheer size is striking. At around 790,000 m2, the Aalsmeer auction is the largest commercially used building in the world ('Largest Commercial Building', n.d.).

Within the industry, flower auctions in general and the one in Aalsmeer in particular are influential:

> *'Events in the [Aalsmeer] auction impact the whole horticultural chain. (…) The auctions determine prices, and those prices reverberate through exporters, wholesalers, and retailers, and Dutch auctions establish a global rubric for floral value. (…) the sales set the standard for price, quality, and aesthetics. Dutch auctions are where you begin to see styles and developments in the industry (…) Dutch flower auctions are an effective barometer of horticultural trade in the Netherlands and around the world' (Gebhardt: 125).*

The flower auction in Aalsmeer combines various economic entities. On the one hand, it is an auction organised as a cooperative. On the other hand, it is the central hub of a global market, connecting the production of cut flowers and the

Fig. 4.4 Dutch Flower Auctions (Locations of Plantation and Royal FloraHolland)

end consumer market. In 2022, 2,814 million rose stems were sold there alone, with a value of almost 700 million euros (Davies 2025b; 2025a). Aalsmeer is not only the largest flower auction but also the oldest, founded in the winter of 1911/ 1912 (Fig. 4.6).

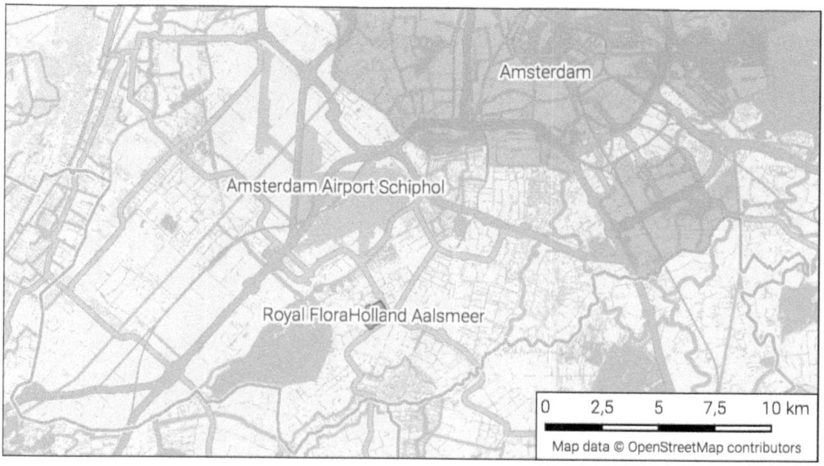

Fig. 4.5 Location of Royal FloraHolland Aalsmeer. (Own illustration, created by Moll)

In addition, both the analysis of the single case study of the commodity chain of Kenyan cut flowers for the German market and the analysis of the global flow of goods show that the Netherlands is central to the (European) cut flower trade, at least in spatial terms. Furthermore, I was able to show that the auction is central to the inherent power structure (2025).

In contradistinction to the financial market, for instance, commodities such as cut flowers are distinguished by a high degree of materiality. The cultivation of cut flowers necessitates months of growth, including years of development, and consequently entails a significant lead time. The storage of cut flowers separated from the mother plant requires sophisticated storage technology, while transport and trade must occur in the shortest possible time due to the high perishability of the goods. These product characteristics shape the commodity chain and exert influence on the inherent power structure. While there are other turnover points and auctions, the flower auction in Aalsmeer is of central importance for the contemporary flower trade. Other auctions, such as those mentioned above, operate more for the regional market and handle different volumes. In view of these aspects, and given the special materiality of the goods and the regionally operating competing auctions, the question arises as to why the flower auction in Aalsmeer as a marketplace and as an actor is so decisive for the cut flower commodity chain today, and how it established and maintained its power.

Fig. 4.6 Inside Royal FloraHolland Aalsmeer (Engel 2022)

In addressing these inquiries, I will undertake a comprehensive examination of the historical evolution of the cut flower commodity chain focusing on the Region Aalsmeer and their flower auction, emphasising its process-nature.

4.2 Data to Examine the Cut Flower Commodity Chain in the Longue Durée

The next essential step, following the case selection and description is the data selection and collection. Both are important components of research in order to answer the research question and be able to generalise from the case. In this book I focus on the structure and process-nature of the cut flower commodity chain.

The flower auction in Aalsmeer is at the centre of the analysis of the cut flower commodity chain. As previous considerations of the research object (Espeter 2025) and the case description above has shown, the flower auction in Aalsmeer is a special actor: it is the oldest flower auction, it is the largest flower auction worldwide and it influences the entire industry in terms of prices and quality standards. The historical reconstruction of the cut flower commodity chain therefore focuses on the flower auction in Aalsmeer as an actor, marketplace and logistical hub. In addition, it acts collectively, representing the interests of flower growers, setting market standards and regulating access (Espeter 2025). There is no comparable actor in this sector.

The historical reconstruction examined in the following is based on the reconstruction and reduction of the cut flower commodity chain from Naivasha, Kenya to Berlin, Germany in the year 2022 (Espeter 2025). Due to the examination of the cut flower commodity chain in the longue durée, a large number of data forms are necessary for the reconstruction of the process-nature. For this purpose, I use data elicited during research periods IV and V (Graeff and Baur 2020) in the form of site visits (Tabacková 2024), walking interviews (Mackay, Nelson, and Perkins 2018; Alexander et al. 2022), and qualitative interviews (Helfferich 2022; Thierbach 2024). These data are complemented by process-produced data (Graeff and Baur 2020): natural (Salheiser 2022) and administrative (Hartmann and Lengerer 2022). In addition, the analysis is based on meta-analyses of pre-interpreted material.

Fieldwork was conducted between March and May 2022. All interviews in the cut flower sector in Nakuru County and at the Nairobi airport, Kenya, were conducted face-to-face, usually at the interviewee's place of work, and lasted between one and two hours. These interviews were supplemented by two further interviews via video call with European managers responsible for cut flowers at FairTrade. For reasons of accessibility, the interviewees were mostly senior managers. The workers' perspective was obtained through walking interviews on individual farms, an interview with the main trade union, and a group interview with a workers' committee on a farm. The firms under study were selected deliberately to represent the widest possible spectrum of the commodity chain, in order to achieve theoretical saturation. Like many other researchers in that field, I struggled during the data collection with limited access to production sites and workers, as also described by Gemählich (2022, 12). This applies to Kenya and especially the Netherlands. For this reason, and for reasons of research ethics, the analysis of pre-interpreted data from periods IV and V is supplemented. The fieldwork around Aalsmeer, the Netherlands, consisted of interviews with logistics and import/export firms, as well as narrative interviews and walking interviews in retail, logistics and on production cities. To cover the retail side, I rely on secondary literature, firm websites, as well as site visits and short interviews in florist shops in Berlin, Germany. German grocery retailers, discounters and online retailers were contacted for interviews, but all declined or did not respond. Unless otherwise stated, all statements in this book refer to data that I have collected and evaluated. An overview of the underlying data material can be found in appendix 1. The collected data is used to ask how the flower auction in Aalsmeer, but also this area, became of such central importance for the cut flower commodity chain. Which processes of establishing and maintaining power can be identified in the longue durée?

To this end, the first step of the analysis is to divide the time under study investigation into meaningful periods.

4.3 Periodisation of the Cut Flower Commodity Chain in the Longue Durée

Using the backward reading method (Baur 2017, 54), I identify the formative phase on the basis of an initial data review. I defined the time under study accordingly. The formative phase 'Imperialism and Curiosity' marks the beginning of my examination of the processes of power establishment and maintenance in the European cut flower market during specific historical periods. This phase describes how cut flowers became a commodity and the market that developed around them, beginning around 1570 with the arrival of tulip bulbs in the Dutch Republic (1579–1795) from the Ottoman Empire and ending in 1850 with the large-scale cultivation of cut flowers for market exchange.

In the following, I use forward reading (Baur, 2017, p. 54) to identify five distinct phases of the cut flower commodity chain. Period I focuses on the early days of commercial cut flower production (1850s–1910s). Period II (1900s–1950s) covers the establishment of central trading centres. From the 1950s to the 00s (period III), economic growth and centralisation became paramount. Phase IV (1980s–2000s) is characterised by changes in the participants and processes involved in the commodity chain. The final phase, Phase V (since the 1990s), is characterised by standardisation within the commodity chain.

The processes involved in establishing and maintaining power in the cut flower commodity chain are summarised and analysed within and along the meaningful periods in Chap. 5. In this analysis I reveal processes that unfold over several periods, focusing on the process-nature of the structures. These are derived from the underlying research question and focus on the processes involved in establishing and maintaining power in the cut flower industry. The event matrix criteria are: (1) processes that take place within or emerge from the structure; and (2) processes of change affecting the structure from outside. Based on these criteria,

I assigned internal and external processes to the identified periods and generated the event matrix (see appendix 2). This was then evaluated in subsequent analysis steps. Using these comparative criteria, I demonstrate how power is established and maintained within the cut flower commodity chain.

The reconstruction and analysis of the structure of actors, spatial structure, and inherent power structures are examined within the periods in Chap. 6. To improve comprehension of the text, this reconstruction is preceded by the final step of the analysis, the examination of the process-nature of the commodity chain in Chap. 5.

4.4 Definition and Operationalisation of Power Within Commodity Chains

Before proceeding the data analysis, I define and operationalise the concept of power, based on the work of Marx (1887 [1867]) and Weber (2019 [1912]), to guide the subsequent analytical steps. My analysis focuses on power relations along the entire commodity chain.

In the modern world-system, power is based on position within the system. It exists in the relationship between actors, where there is both political and economic power (Wallerstein 2011a [1976]; 2006 [2004]). Therefore, I argue, in economic structures such as commodity chains, power is determined by one's position within the system. Within these structures, a distinction is made between political and economic power.

While both Marx and Wallerstein (Marx and Engels 2015 [1848]; Marx 1887 [1867]; Wallerstein 2011a [1976]) emphasise the interplay between political and economic power, neither provides a comprehensive definition. Therefore, I use Weber's definition of power in the empirical analysis. Weber (2019 [1912]) describes power as 'jede Chance, innerhalb einer sozialen Beziehung den eigenen Willen auch gegen Widerstreben durchzusetzen, gleichviel worauf diese Chance beruht' (Weber 1980 [1921], p. 28), meaning: every opportunity to assert one's will in a social relationship, even in the face of resistance, regardless of the basis of this opportunity. To make this concept analytically useful for commodity chains, I specify it in terms of political and economic power.

In order to examine the inherent power structure, it is necessary to operationalise the parameters of political and economic power. This operationalisation can be applied to economic structures such as commodity chains. Therefore, I define economic power in the sense of Marx and Wallerstein as access to and control over resources and means of production. On this basis, this power can manifest itself in the form of financial resources, but also through influence on pricing and thus on the generation of added value along the chain. Furthermore, I define political power as the ability to regulate processes, especially by state and supranational actors, and to influence political decision-making. This regulation can be expressed, for example, through economic policy measures such as steering mechanisms (taxes, subsidies), legislation or even passivity (Wallerstein 2006). This regulation of processes can affect a wide range of steps and actors along the chain. Political power, which in turn primarily affects production, is the restriction of market access. This is usually done by state and supranational actors in the form of protectionist measures (again, influence on political decisions). However, non-state actors can also influence and restrict market access.

Part II
Empirical Study

5 Process-Nature of the Cut Flowers Commodity Chain

'Equally, if not more important, the examination of a commodity chain over time allows the observer to assess the nature and degree of structural transformations of the organization of the chain' (Hopkins and Wallerstein 1986, 162).

In the following chapter, my analysis of the cut flower commodity chain will focus on the processes of establishing and maintaining power within the chain. During the research process, this step was carried out by analysing the commodity chain itself. For the purpose of clarity, I have reversed the order of these steps in the analysis. In subsequent Chap. 6, I will reconstruct the periods identified in the analytical process of periodisation. The focus is on the cut flower commodity chain as structure of actors, spatial structure and inherent power structure within the meaningful periods.

In the following chapter, I analyse the process of establishing and maintaining power in the cut flower commodity chain along meaningful periods. Starting from the formative phase, I divide the periodisation into content-based, meaningful, periods, as described in detail in Sect. 4.3 and illustrated in Fig. 5.1.

By examining each period and the periods that follow, I reconstruct the internal and external processes that have led to the current structure of the commodity chain. This allows me to examine the central importance and position of the Aalsmeer flower auction in the cut flower industry.

L. M. Espeter, *From Hobby to Billion-Dollar Industry*, Wirtschaft + Gesellschaft,
https://doi.org/10.1007/978-3-658-52437-1_5

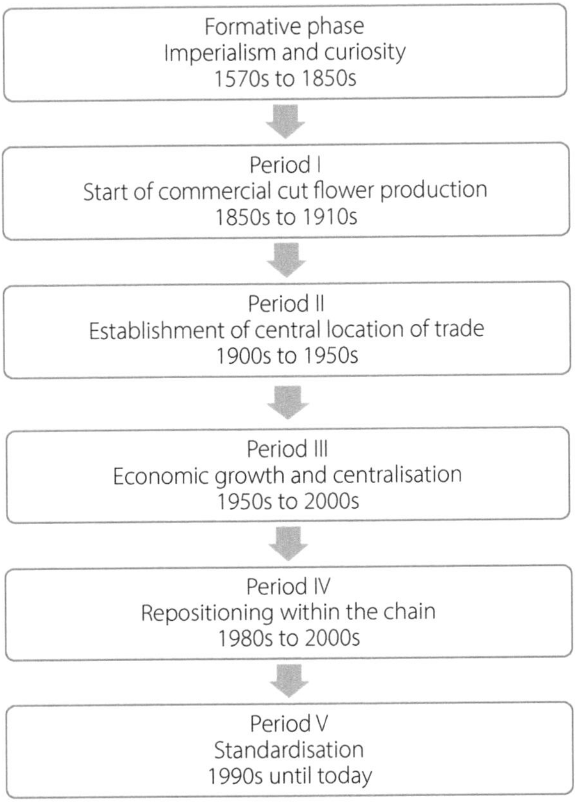

Fig. 5.1 Periodisation of the time under study. (Own illustration)

5.1 Formative Phase: Imperialism and Curiosity (1570s to 1850s)

My analysis of the formative phase of the Dutch cut flower industry is based on a metadata analysis of historical literature and natural data sources. In the following, I describe how tulips and other bulbous plants arrived in the Netherlands. I examine how they initially became the subject of research by a single natural scientist and an interested group. I discuss how growing interest in unusual flower shapes and colours probably caused the first speculative bubble of its kind, and how cut flowers eventually became a sought-after product thanks to technological developments that simplified their cultivation.

Even though flower sellers existed in ancient times — for example, in Greece and the Roman Empire (Goody 1993, 57ff) — and the commodification of cut flowers was already taking place, this book focuses on the modern cut flower industry. More specifically, I will examine the area that is now the Netherlands, paying particular attention to the Aalsmeer region in North Holland.

During the 17th century, the United Netherlands (also known as the Republic of the Seven United Netherlands, 1579–1795) experienced its heyday. In this period, political and economic structures were established that would later contribute to the development of the cut flower industry. The Republic gained independence from the Spanish monarchy during the Dutch Revolt in the late 16th century, a status that was formally recognised in the 'Treaty of Münster' in 1648 (Kennedy 2017, 182). It consisted of seven autonomous provinces (van Heck 2021, 24). Characterised by a Calvinist society which also granted religious freedom, power was primarily in the hands of merchants and financiers in the cities and provinces (van Heck 2021, 24; Gebhardt 2014, 54). Unlike other European societies, whose politics and economies were still heavily influenced by the nobility (Gebhardt 2014, 53), the Dutch Republic attracted capital and trade interests from across Europe (van Heck 2021, 24).

The expansion of the Dutch sphere of influence through imperialism, including by the trading company 'Vereenigde Oostindische Compagnie' (VOC) (English: 'United East India Company'), had a profound impact on Dutch society. The emergence of consumer culture and the development of capitalism in the Netherlands would probably not have been possible without imperialism (Gebhardt 2014, 55). The VOC, an association of Dutch merchants, was granted a trade monopoly and sovereign rights in conquered territories by the Dutch crown. This enabled the VOC to become one of the largest trading companies of the 17th and 18th centuries. Its network of merchants, already central to the Netherlands, now extended globally. Port-cities such as Middelburg became important transshipment points for goods, animals, and plants from the colonies. These exotic new goods were particularly popular among the Dutch upper classes, as evidenced by the group of 'liefhebbers' (in English: enthusiasts). Some of these plant enthusiasts were early investors in the VOC.

The VOC played a significant role in the emergence of the Dutch flower industry. Among other things, the VOC was mandated to search for new plants and flowers, and it maintained close ties between emerging Dutch horticultural art and the national colonial project (Gebhardt 2014, 74). Port-cities such as Middelburg, an important VOC base, contributed significantly to the development of Dutch horticulture. Middelburg quickly became a centre for floristry (Goldgar 2007).

The economically flourishing Netherlands of the 17th century was characterised by a spirit of trade and colonial connections. This historical context created important conditions for the growing interest in and trade of flowers, which later developed into a significant economic sector.

The cut flower industry has its origins in flower bulbs. At the beginning of the 17th century, interest in flower bulbs was growing, particularly in the Netherlands. Initially, this interest was confined to a select group of wealthy enthusiasts, known as 'liefhebbers', who exchanged information about plants and flower bulbs, as well as their cultivation. This group gave rise to a market for trading bulbs, which spread throughout the Netherlands. Carolus Clusius was central to this network of enthusiasts. Born in the Spanish Netherlands in 1526, Clusius was a botanist who was appointed head of the garden at Leiden University in 1593. He acquired extensive knowledge of plants through extensive travel and his role as Prefect of the Imperial Medical Garden in Vienna, among other things. He shared his knowledge with countless contacts and documented it in correspondence that has since been compiled and digitised in the Clusius Project ('Clusius Correspondence' 2015). Clusius himself was an enthusiastic collector and grower of tulips. At his request, his friend Ogier Ghiselin de Busbecq — a botanist who was ambassador to the Ottoman Empire at the time — sent him flower bulbs and seeds (van Heck 2021, 27). Upon taking up his position at Leiden University, Clusis established a hortus botanicus—the first in the Netherlands—rather than a hortus medicus (van Heck 2021, 52). He placed particular emphasis on tulips (van Heck 2021, 29).

> *'Clusius and his network of hundreds of men and women all over Europe helped to shape knowledge through the careful and meticulous exchange of information, via letters, about varieties, growing conditions (soil, water, and weather), medical usage of plants, or about the beauty and color of flowers. Based on this 'collective enterprise,' the academic field of botany and the flower bulb industry were established' (van Heck 2021, 27).*

The network of flower enthusiasts comprised men and women from the upper middle classes, including merchants, doctors, lawyers, politicians and bankers. They had connections to various enterprises, such as the VOC and the 'Beemster Polder'. Together with Clusius, this group significantly changed the image of Dutch gardens. Where medicinal or agricultural plants had once been cultivated, gardens for aesthetic purposes were now created. This was true both within and outside cities, as maps of Haarlem from 1649 show (Goldgar 2007, 48–49). However, the flower enthusiasts were not only fascinated by the aesthetics of these new plants. They also investigated their potential medicinal uses and genetics. In particular, they were interested in the colouring and 'breaking' of the flowers, and wanted to understand their inheritance (Goldgar 2007; Gebhardt 2014; van Heck 2021). In doing so, they laid the foundation for commercial flower growers. The first nurseries were established near Haarlem as early as the beginning of the 17th century, as the sandy soil behind the dunes was ideal for growing bulbs.

By the mid-17th century, a thriving trade in flower bulbs had already emerged, particularly in the Netherlands. This trade usually took place at auctions, which 'were typically held in taverns and in other public places' (van Heck 2021, 77). The auction determined the price and buyer (the highest bidder) of the products in question. As flower bulbs are only available at a certain time of year, these auctions were mostly contracting for the future. They are planted in September, left in the ground over winter, and harvested in summer after blooming in spring. However, at that time, contracts for future transactions were prohibited in the Netherlands (van Heck 2021, 78f). This meant that the seller and the buyer had to trust each other implicitly.

In retrospection, the early heyday of the flower bulb trade is often referred to as 'tulip mania' It was the first speculative bubble. Prices rose rapidly in the 1630s, especially for special tulip bulbs, before collapsing in 1637. A similar trend occurred for hyacinths a little later. Historian Anne Goldgar's analysis (2007) this period based on transaction documents, as well as letters and other sources. Her study shows that fewer people were involved in the specialised tulip trade than often assumed. Furthermore, while some people lost money, no one went bankrupt. The overall impact on the Dutch economy was minimal. Additionally, Goldgar examines that the trade of flower bulbs largely occurred within close personal networks, such as among relatives and within religious groups.

After the establishment of flower cultivation, a market for cut flowers developed, influenced by various far-reaching changes. Initially, cut flowers were a highly exclusive luxury item, restricted to the privileged classes. As previously mentioned, the expansion of European influence through imperialism played a decisive role in the emergence of the cut flower commodity chain. This brought exotic products to the Europe. Those who could afford it collected rare fossils, colourful butterflies, stuffed birds and other animals from faraway lands (van Lier 2005, 17). Additionally, the French Revolution (1789–1799) profoundly impacted Europe as a whole, including the Netherlands. The occupation of the Netherlands and the establishment of the Batavian Republic in 1795 resulted in political changes based on the French model, including the implementation of a centralised government (Wintle 2000, 156–57; Kennedy 2017, 261ff). The French Revolution also impacted the social fabric throughout Europe, leading to a stronger desire for social equality. A climate emerged in which social hierarchies and exclusivity were increasingly questioned in some circles. This was reflected in consumer behaviour, for example, with an increased demand for goods such as cut flowers (van Lier 2005, 20).

It can be assumed that the sales market for cut flowers emerged because the product was already being traded as tulip bulbs, meaning it was already well known. Additionally, tulip bulbs posed a problem due to uncertainty surrounding their appearance (Goldgar 2007, 197ff), and were likely unappealing to the general market as an investment opportunity.

Over the decades and centuries, interest in the production and sale of tulip bulbs grew (van Heck 2021, 37). In the 18th century, Dutch tulips were already internationally renowned for their quality and variety. The Netherlands exported flower bulbs to Germany, France and England, among other places (Gebhardt 2014, 92). By the early 19th century, trade and commerce of bulbs had become a respected profession in Dutch society. People could make money and gain social prestige: '[t]he trade of tulips became a logical component of that economy and society' (van Heck 2021, 26). During the 19th century, the cut flower market became so specialised that the first recorded florists (Dutch: bloemisten) appeared in Haarlem in 1802 and in Utrecht and other Dutch cities in 1855 (*De Dordrechtsche Courant* 1802; *De Dordrechtsche Courant* 1855).

Over time, Dutch flower growers and traders gained a reputation throughout Europe, and their products became highly popular. It was common knowledge that the finest bulbs came from the Netherlands. Although there were also some important nurseries in Belgium and France, Holland continued to have the greatest expertise. The flower bulbs were primarily produced for export, but also for use as cut flowers domestically (van Lier 2005, 20). Ideal conditions for producing flower bulbs were found in the region between Leiden and Haarlem. They thrived in the open air of the coastal climate and the well-drained, sandy soil (van Heck 2021, 37; Gebhardt 2014, 92).

The situation was different for other plants, which were more or less tropical and did not thrive in the region's climatic conditions. The first greenhouse designed specifically for tropical plants was built in a botanical garden around 1680. These structures protected the plants from external environmental influences and created an optimal environment. Glass greenhouses enabled sunlight, temperature and humidity to be regulated and controlled (van Heck 2021, 53ff). They also 'protect crops, increase productivity, and make more efficient use of water, light, energy, and fertilizer. When a crop is sheltered, the whole internal climate will differ from outside conditions' (van Heck 2021, 56). Around 1850, the first greenhouses were used in the Netherlands for growing grapes. Around fifty years later, these greenhouses could be heated, enabling the cultivation of more sensitive plants and other products. Consequently, flower growers could now cultivate lilacs and roses in greenhouses. This marked the beginning of cultivating cut flowers that did not come from bulbs at the end of the 19th century (van Lier 2005, 20–21).

In addition, the usable area of the Netherlands increased significantly in the 17th and 18th centuries due to drainage projects. The first of these projects was the 'Beemster Polder' in 1612, which is now a UNESCO World Heritage Site (UNESCO World Heritage Centre, n.d.). Investors in the project included 'liefhebbers' (Gebhardt 2014, 62ff). This type of land development had a significant impact on the Dutch people's self-image: '[a]n important part of the Dutch imaginary involves reshaping the landscape, creating arable land where before there was swamp and bog, or lake and sea' (Gebhardt 2014, 63). The complete drainage of the Haarlemmermeer in the mid-19th century was a significant factor in the emergence of commercial cut flower production in the Aalsmeer region (Fig. 5.2). Until then, fishing in the Haarlemmermeer had been a key source of income for the inhabitants of Aalsmeer, as the town bordered the lake (*Schilder's Niewsblad* 1967). The lake was drained in 1837 in response to storm surges (*De Dordrechtsche Courant* 1836) and the threat of flooding in Haarlem, Leiden and Amsterdam (*De Dordrechtsche Courant* 1837). The work began in 1842 (*De Dordrechtsche Courant* 1842). In 1852, the lake was completely drained (Wintle 2000, 113). Initially, the population of Aalsmeer cultivated vegetables in the very peaty soil (van Lier 2005, 27). Shortly afterwards, the cultivation of flowers began, increasing significantly from the 1890 s onwards (*Schilder's Niewsblad* 1967).

In summary, a variety of factors interacted over a long period of time to establish the commercial production of cut flowers in the Aalsmeer region. First, there was a fascination with exotic plants in Dutch society during its cultural and financial peak in the 17th century. During this period, imperialism and seafaring gave rise to a large, well-connected social class of merchants and financiers. Conversely, the French Revolution (1789–1799) and its aftermath brought about changes to societies throughout Europe. Prosperity grew among large sections of society, as did the interest in and ability to afford things such as cut flowers. Furthermore, Gebhardt concludes that changing ideas about masculinity and femininity during this period influenced the way in which flowers were cultivated, meaning they had to serve aesthetic *and* commercial purposes. This promoted the commercialisation of flower cultivation (Gebhardt 2014, 93). Technological

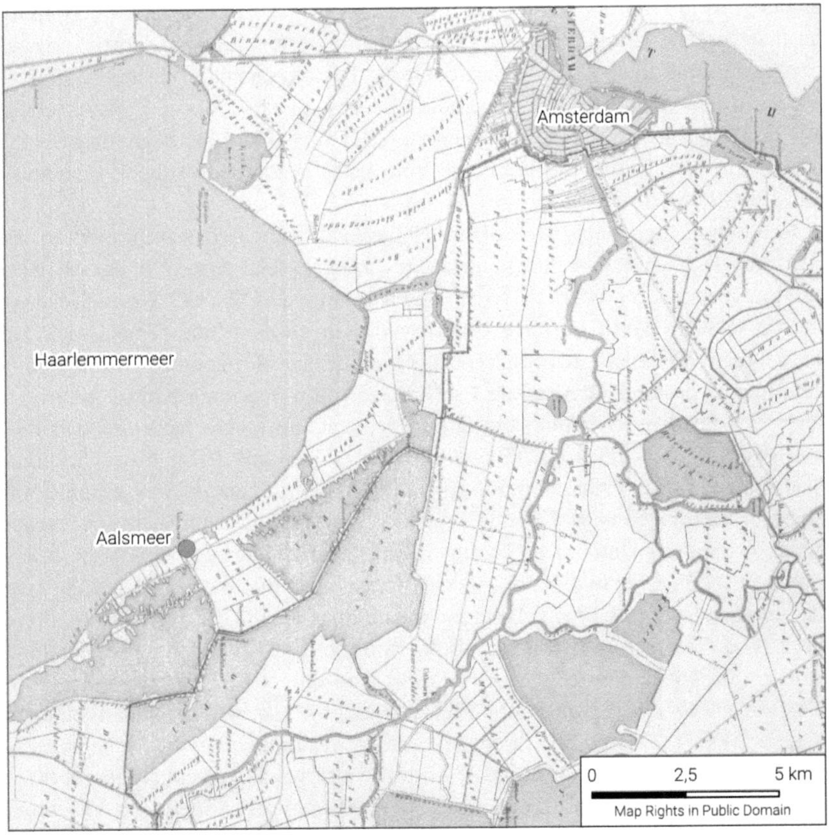

Fig. 5.2 Location of Aalsmeer 1860 (Own illustration, coloured by Moll)

innovations such as greenhouses and steam engines, as well as large construction projects such as the 'Beemster Polder' and the drainage of the Haarlemmermeer, made the commercial cultivation of cut flowers possible and necessary in the Aalsmeer region as a new source of income. Therefore, the spatial expansion of the Netherlands was crucial in two respects during the formative phase of the cut flower commodity chain. First, imperialism, including by the VOC, sparked society's interest in new and exotic things brought to the Netherlands. At the same time, Dutch colonisation resulted in the emergence of a large and influential social class of merchants and financiers, who amassed capital through the exploitation of violently appropriated regions. Second, the land area of the Netherlands increased as a result of drainage projects. These different forms of spatial expansion made new areas accessible for a new form of capital accumulation.

5.2 Period I: Start of Commercial Cut Flower Production (1850s to 1910s)

Around 1860, there were just under a hundred flower growers in Aalsmeer (van Lier 2005, 27). In the early days of the cut flower industry, farms distributed their flowers themselves. The farmers in Aalsmeer specialised in growing lilies and roses (Gebhardt 2014, 93), initially cultivating cut flowers alongside vegetables. By 1880, Aalsmeer had developed into a centre for flower cultivation and nine out of ten of the barges at Amsterdam's flower market came from Aalsmeer (van Lier 2005, 27).

At the end of the 19th century (1886), the number of heated greenhouses used for flower cultivation in Aalsmeer grew from one to more than 200 within twelve years (*Schuitemakers Purmerender Courant* 1886). Shortly afterwards, in 1896, the Aalsmeer branch of the Royal Dutch Society for Horticulture and Botany (Koninklijke Nederlandse Maatschappij voor Tuinbouw en Plantkunde) applied to open a horticultural school in Aalsmeer. The request was apparently granted immediately, as the first class at the new state secondary school for floriculture in Aalsmeer, the 'Rijks Middelbare Tuinbouwschool voor Bloementeelt te Aalsmeer' began in 1897 (Nationaalarchief 2025b). This was the only school in the Netherlands focusing specifically on floriculture; other horticultural schools, by contrast, focused on horticulture in general (*Betuws(Ch) Tuinbouwblad* 1965a).

At the same time, increased cultivation of cut flowers made them affordable for the mass market, leading to growth in sales (Gebhardt 2014, 93). This, subsequently, led to the emergence of a professionalised trade in cut flowers. Traders sourced their goods directly from growers and sold them at markets to consumers or to the emerging floriculture industry. Examining the power structure of period I—which I conduct in 6.1.4.—reveals that farmers are at a disadvantage compared to traders. Unlike traders, farmers are in direct competition with each other, meaning they have little influence over pricing (Gebhardt 2014, 93; van Lier 2005, 21–22). Furthermore, merchants did not always pay their bills reliably (van Lier 2005, 21–22)., putting additional economic pressure on farmers. Additionally, the first Flower Trade Organisation, the 'Nederlandse Tuinbouwraad' was founded in 1908 as a merchant interest group (Nationaalarchief 2025a). These processes led to retailers strengthening and consolidating their position of power in terms of economic and political power.

Furthermore, during period I, state actors continued to create land for agricultural use through land reclamation projects (Wintle 2000, 113). In addition, public infrastructure projects were implemented, including the construction of canals and the expansion of the rail and road networks (Wintle 2000, 104). These projects benefited the growing cut flower industry in particular, as the canals were largely used to transport coal for heating greenhouses and to transport cut flowers to sales markets (Porter, Ramirez-Vallejo, and van Eenennaam 2013, 6).

5.3 Period II: Establishment of Central Locations of Trade (1900s to 1950s)

The processes that led from period I to period II can be traced back to the growing imbalance of power between traders and farmers. Cut flower growers in Aalsmeer reacted differently to their deteriorating position in the power structure. In 1899, some of them established 'De Vereenigde Tuinbouwers' (engl.: United Horticulturalists), the first farmer-run cut flower market in Amsterdam (Gebhardt 2014, 93). Conversely, the first discussions about setting up flower auctions in Aalsmeer took place in 1910 (burocontent 2021). In the winter of 1911/1912, the first farmers came together in Aalsmeer to sell their goods collectively, close to where they were produced (Gebhardt 2014, 93). In December 1911, the cooperative Bloemenlust was founded in Café Welcome. Five days later, on 5 January 1912, the first flower auction of the Centrale Aalsmeerse Veilingvereeniging (CAV) took place at the Drie Kolommen (van Heck 2021, 90–91).

In addition to the power imbalance between production and trade, however, van Heck also identifies the growing market as a reason for the need to professionalise the cut flower industry: 'Most growers preferred to concentrate on the production of flowers; the shippers were too busy with transport and reaching the cities to sell the produce to retailers and local markets. Buyers were not fully trusted to handle the auction process' (van Heck 2021, 90). This assumption is supported by the emergence of firms and producers specialising in individual steps, such as breeding, along the entire commodity chain during this period (Gebhardt 2014, 95).

Furthermore, the establishment of auction cooperatives, such as flower auctions, was influenced by external auction legislation and the associated cooperative movement in the Netherlands. At the beginning of the 20th century, there was a rise in cooperative foundations throughout the Netherlands:

> *'Individuals and communities began to understand that they could protect themselves to some extent from an exploitative system by forming collective organizations. (...) rural cooperative banking (...) Fruit and vegetable auctions, and slightly later the Aalsmeer flower auctions, were founded in this milieu of agricultural cooperatives, industrialization, and a sense of egalitarianism' (Gebhardt 2014, 28).*

The establishment of the flower auctions was influenced by external factors such as auction law, the cooperative movement and the first horticultural auctions at the beginning of the 20th century (Meulenberg 1989, 147–48). These auctions have always served the same function: 'price discovery, quality control, assembling of produce, and warranty of payment to growers' (Meulenberg 1989, 148). Over the next five years, more flower auctions were established in Rijnsburg and other areas of Westland. 'But from the beginning, in many ways Aalsmeer led the industry. [...] [I]t had not only the largest market share, but also excelled with the best established export position' (Gebhardt 2014, 95–96).

Through cooperatively organised auctions, flower farmers gain collective power (see Arendt 1970). On the one hand, they now regulate market access by making cooperative membership compulsory for producers (political power). At the same time, the political power of the auction is evident in its members being obliged to sell their goods via the cooperative's own marketplace. This minimises the possibility of goods being sold bypassing the auction. In turn, the traders must participate in the auction to purchase goods (van Lier 2005, 12). Collectivising their interests gave the farmers considerable advantages. This significantly improved their bargaining power with regard to pricing. The establishment of the cooperatives and their subsequent consolidation into the 'Kring van Bloemenveilingen in Nederland' (engl. Circle of Flower Auctions) in 1933 opened up new financial opportunities for flower growers (*Utrechtsch Nieuwsblad* 1933b). They were now able to make joint investments (van Lier 2005, 12). At that time, there were flower auctions in Aalsmeer, Utrecht, Boskoop, Nijmegen and Hilversum (*Utrechtsch Nieuwsblad* 1933b).

The organisational structure of flower auctions shapes the entire commodity chain. Consequently, '[o]ther aspects of the chain, like breeding, wholesaling, and retailing, have fallen into place around the cooperative auction structure' (Gebhardt 2014, 95). Moreover, interest groups for flower auctions were established in period II.

The political power of auction cooperatives increased during period II due to the establishment of additional foundations and consistent membership growth. This is because bypassing the cooperative trading centres is difficult. The importance of the cut flower sector in the Netherlands is demonstrated by external influences on it, among other things. During this period, government actors increasingly took up the interests of the cut flower commodity chain (political power—influence/lobbying). For example, this occurred during times of external shocks, such as the Great Depression in the Netherlands (1933–1936). The global economic crisis halved auction sales (Royal FloraHolland. n.d.). During this period, the Netherlands passed an agricultural subsidy law (1933), which included tax relief and loans for flower growers in Aalsmeer and Haarlemmermeer. The support was justified as follows: "in that region, flower cultivation plays such an important role in the economy that its decline in that area would have had disastrous consequences for the entire population of that region' (Deventer Dagblad 1933—translated from the Dutch original by the author).

The specific support given to this region was questioned and criticised by flower producers outside this region (*Utrechtsch Nieuwsblad* 1933a)

The economic crisis of the 1930 s was followed by the Second World War (1939–45) and the subsequent occupation of the Netherlands by Nazi Germany (1940–45). This period had a particularly far-reaching impact on the cut flower industry at the time. There was no government support, quite the opposite in fact. Prior to the Second World War, the Netherlands primarily exported flowers to neighbouring countries such as Belgium, Luxembourg, Germany, England, Scandinavia, and Switzerland. During the war years, these exports declined completely (van Lier 2005, 23). Germany had previously been the main export destination for cut flowers sold at the Aalsmeer auctions. At the start of the war, sales at the Aalsmeer auctions declined dramatically. In the first half of 1940, combined sales at the Centrale Aalsmeerse Veiling and Bloemenlust totalled 2,654,129 guilders. This represented a 20% decline compared to the first half of 1939 (*De Zaanlander* 1940). There were no flower markets during the occupation of the Netherlands by the German Wehrmacht. Furthermore, in 1942, the auctioning of lilacs, carnations and roses was prohibited. The given reason was the high energy requirements of the greenhouses (Gebhardt 2014, 104).

5.4 Period III: Economic Growth and Centralisation (1950s to 2000s)

The post-war period following the Second World War marks the transition from period II to period III. This time was characterised by both resource scarcity and general economic growth.

During the post-war reconstruction period, the Netherlands swiftly re-established itself as a leading exporter of cut flowers. Although there was competition from France and Italy, the Netherlands remained the leading exporter. Due to booming demand, primarily from West Germany but also from within the Netherlands, the sector experienced enormous growth in the 1950 s and 1960 s (van Lier 2005, 23–24). General economic growth influenced the cut flower sector. The entire industry benefited from the recovering sales market, which grew gradually (van Heck 2021, 91). Furthermore, the Treaty of Rome, which established the European Economic Community in 1957, facilitated access to European sales markets by creating a common internal market between the Netherlands, Belgium, France, Luxembourg, Italy and Germany (European Parliament 2025). In addition, the entire industry benefited from its influence (political power):

> *'On July 1, 1968, a European free market for flowers and plants was established. The system of cultivation permits was abolished in 1969, leading to lower barriers to entry to start as a grower or exporter. The economic results were spectacular. Exports from the Netherlands to other European countries increased threefold and the Dutch began to expand exports beyond adjacent countries like Belgium, Germany, and Luxemburg to reach France and Italy. The increased demand for flowers and plants' (van Heck 2021, 91).*

During the remainder of period III, the export area expanded with the accession of Spain and Greece to the European Community in 1986 (European Union, n.d.-b; n.d.-c). This was followed by the fall of the Iron Curtain in 1990 (Britannica 2025), after which Eastern European countries gradually integrated into the European internal market. In 1990, the flower auction in Aalsmeer achieved a turnover of two billion guilders (van Lier 2005, 43).

The growth of the sales market in the 1960 s also marked the transition from period II to period III, with the CAV and Bloemenlust cooperatives in Aalsmeer deciding to merge as the Cooperative *Society* 'Verenigde Bloemenveilingen Aalsmeer' (VBA) in 1964 (Gebhardt 2014, 97). The rapid growth of the Aalsmeer auctions after the Second World War put particular pressure on the CAV in the mid-1960 s. Located at Van Cleeffkade between the village and the nurseries, the auction had no real possibility of further growth. The only viable option for expansion was to relocate (van Lier 2005, 30). At the same time, both the CAV and Bloemenlust had to defend their positions within the structure and secure their power. The historically established spatial division of Aalsmeer was challenged by new flower farmers. Furthermore, the market shares of the CAV and Bloemenlust were reduced by the establishment of new flower auctions in Westland, Beverwijk, and Vleuten (van Lier 2005, 30). The flower auction in Aalsmeer initiated a merger in 1964, commissioning Reifeisenbank to mediate between the two parties. The product ranges of the two marketplaces complemented each other: 'at Bloemenlust, only flowers (mainly roses and carnations) were auctioned, while at CAV, more lilacs and tulips, as well as plants, were auctioned' (van Lier 2005, 30–31). The merger of the CAV and Bloemenlust also brought social and religious groups in Aalsmeer together. Additionally, the spatial division of the town was overcome.

Several potential locations for the new auction were considered during the planning process. Even moving to Haarlemmermeer was considered. However, the municipality of Aalsmeer successfully pursued its economic interests by keeping the auction within the municipal boundaries (see Fig. 5.3). The site offered by the municipality had sufficient space for the auction, both now and for future expansion. In addition, this location has excellent transport links. The first stone was laid for the new building for the merged VBA on 29 May 1970 (van Lier 2005, 32–33). It opened in 1972 and has been the world's largest industrial building ever since ('Largest Commercial Building', n.d.). Due to the continuing growth of the sales market, the flower auction expanded shortly afterwards during the uncertain times of the 1970s oil crisis (external shock). By 1975, it had finally expanded to cover 150,000 m2 (van Lier 2005, 38).

The spatial centralisation is evident in the expansion of the auction in Aalsmeer. Furthermore, more and more other stakeholders settled around the new VBA location, including the 'Nederlandse Tuinbouwraad' (engl.: Dutch Horticulture Council) (Nationaalarchief 2025a), trading and logistics companies (Porter, Ramirez-Vallejo, and van Eenennaam 2013, 6; van Lier 2005, 34). In the 1980 s, the flower auction in Aalsmeer expanded again, developing adjacent areas (van Lier 2005, 42). In the 1990 s, the steadily increasing volume of traffic around the auction site caused problems. As a result, the area was expanded once more. The Aalsmeer flower auction expanded southwards across the Legmeerdijk (VBA-Zuid, later FloraHolland-Zuid) (van Lier 2005, 47). The newly developed site was leased to traders and other stakeholders, who relocated their businesses and warehouses there. These areas are connected to the auction by an overhead conveyor system (Logistic Expert Royal FloraHolland and Docservice Manager 2022). By 2004, the area covered by the Aalsmeer flower auction was approximately one million square metres, 'more than ten times the area occupied by the merger partners in 1972' (van Lier 2005, 47).

The expansion of the Aalsmeer flower auction as a central place in the commodity chain was supported by the state. For instance, the auction received state funding to finance its new building in the 1970 s. The construction of the building, which cost a total of 57 million guilders, was financially supported by the EC (6.25 million guilders) and the Dutch government (1 million guilders) (van Lier

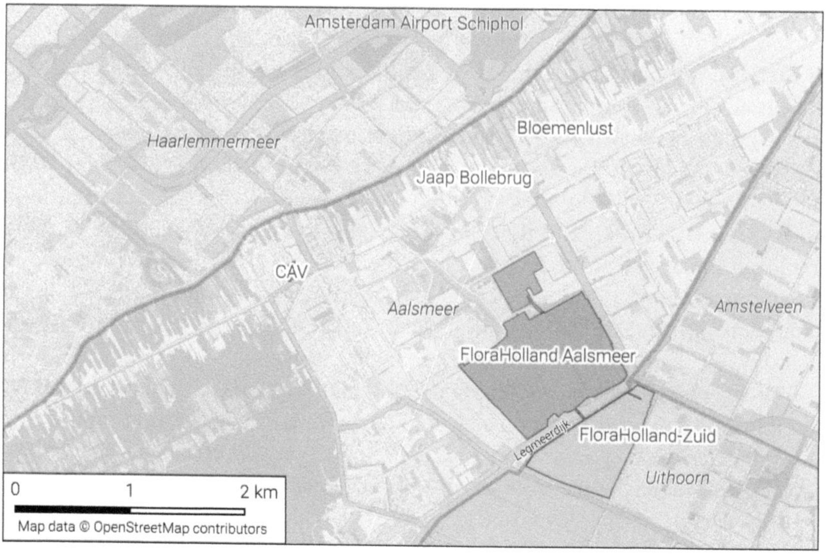

Fig. 5.3 Geographical position of the flower auctions in Aalsmeer (Own illustration, created by Moll)

2005, 34). In addition, the surrounding infrastructure was expanded and financed by the public sector:

> *'In either case, 'Aalsmeer's auctions have been integral to regional planning over the past century. When building highways, intersections, traffic lights and so on, planners have had to take into account the route and schedule of thousands of trucks coming from and going to the auction. Such volumes of traffic also affect quality of life issues for local residents' (Gebhardt 2014, 97).*

In addition, the flower auction in Aalsmeer benefits from the 'Mainport Strategy' during period III. Schiphol Airport, which was formerly used for military purposes, is located on the reclaimed Haarlemmermeer (see Fig. 5.3). After the war, the Dutch government invested in its restoration so that Schiphol, therefore the Netherlands, could establish an important role in global trade (Huijs 2011). Supporting important transhipment points such as the Port of Rotterdam and Schiphol Airport as 'key drivers of the Dutch economy' was part of the national policy agenda known as the 'Mainport Strategy'. The aim was to establish the Netherlands as a distribution hub—'Nederland Distributieland' (OECD 2014, 117). To this end, Schiphol was significantly expanded, partly against the wishes of local residents (*NRC Handelsblad* 1993; *Algemeen Dagblad* 1995).

The discovery and development of gas fields in northern Netherlands had also influenced the cut flower commodity chain from external. This has provided cut flower producers and their suppliers with access to cheap energy, among other advantages (Ministerie van Klimaat en Groene Groei, n.d.; vaan Oostrom 2011, 108–9; *Reuters* 2024). In addition, the industry relied on illegal and semi-legal workers. From the 1970s to the 1990s, they accounted for around 20% of those employed in cut flower production: 'In the early 1980s some growers were complaining that the only way they could earn a profit was to hire underpaid illegal immigrant workers, which they did in spite of the 10,000 guilder fine introduced in 1979' (Gebhardt 2014, 105).

The EU's enlargement in 2004 and 2007, and the free movement of workers that followed seven years later, had a significant impact on cut flower production. Since then, as in other areas of agricultural production, workers in cut flower production have mainly come from eastern EU regions. The wages of these workers, most of whom are employed on a seasonal basis, are significantly lower than those of Dutch employees (Trade Union Official (NL, EU, International) 2022). Lower fixed costs improve the economic position of Dutch producers and the auction, where many Eastern Europeans are also employed (Trade Union Official (NL, EU, International) 2022). The centralisation of market, capital and profit power through the merger of auction cooperatives has improved the position of the marketplace, the cooperative and its members within the power structure of the cut flower sector (politically and economically).

Following the initial wave of mergers, the remaining flower auctions established the Dutch Flower Auction Association (VBN) in 1974 to serve as an umbrella organisation. The VBN acts as a lobbying organisation for the auctions, giving them political power (Vereniging Bloemenveiling Nederland, n.d.). In addition to setting political agendas, the VBN provides a wide range of information on product specifications, thereby shaping market access and political power. This includes images showing the stages of maturity of a variety of cut flowers, for example. This classification of maturity stages is one of the pieces of information provided to the trade and influences pricing (Auctioneer 2022).

In 1945, there were seventeen flower auctions in the Netherlands (Meulenberg 1989). Centralisation became apparent from 1985 onwards (see Fig. 5.4). Of the twelve auctions in 1985, only six remained by 1995 (Gebhardt 2014, 29–30), and just four by 2002 (Porter, Ramirez-Vallejo, and van Eenennaam 2013, 7). In 2005, VBA Aalsmeer—'by far the largest' (Porter, Ramirez-Vallejo, and van Eenennaam 2013, 7)—and FloraHolland merged. Since 2008, the Oost-Nederland and Vleuten flower auctions have merged under the name Plantation. Following the merger of the VBA and FloraHolland in 2008, FloraHolland (with locations in Aalsmeer, Naaldwijk, Rijnsburg, Venlo, Bleiswijk and Eelde)—since 2011 Royal FloraHolland—and Plantion (in Bemmel and, since 2010, Ede) are the only remaining flower auctions in the Netherlands (Royal FloraHolland, n.d.; Plantion, n.d.).

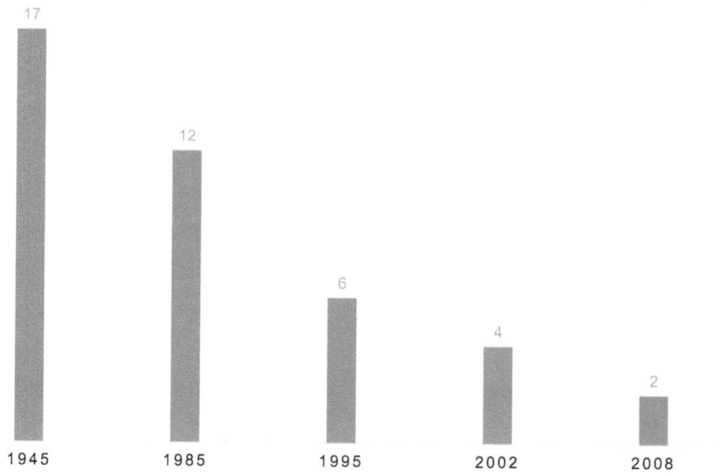

Fig. 5.4 Number of cooperatives flower auctions over time (Own illustration, based on Meulenberg 1990, Porter, Ramirez-Vallejo, and van Eenennaam 2013, and Gebhardt 2014)

A merger between FloraHolland and Plantation has so far been rejected by the latter (Gebhardt 2014, 102). Efforts were made to take action under antitrust law against the 2008 merger of VBA and FloraHolland, which created a quasi-monopoly in the cut flower commodity chain. However, the Board of the Netherlands Competition Authority rejected this on the grounds that: '[t]he Executive Board is of the opinion that the proposed concentration will not lead to the creation of a dominant position or a strengthening of a dominant position that would significantly impede effective competition on the Dutch market or part thereof' (NMa 2007, 35—translated from the Dutch original by the author).

The decision was presumably made because the VBA and FloraHolland were already dominant actors. Shortly before the merger in 1990, they accounted for 75% of auction sales (Meulenberg 1989, 153).

The centralisation of flower auctions in Aalsmeer has led to the emergence of a commodity chain focused on this location:

> *'If there is one place where the power of cooperative collaboration has been proven, it is in Aalsmeer, where the merger of two cooperative auctions marked the beginning of the creation of a world trade centre for the floriculture industry on an unprecedented scale. For sixty years, the Bloemenlust and Centrale Aalsmeerse Veiling (CAV) auction cooperatives grew at the same pace, which was impressive in percentage terms but modest in absolute figures. When the two auctions joined forces in 1972 to form Verenigde Bloemenveilingen Aalsmeer (VBA), Aalsmeer became the undisputed centre of the flower trade in the Netherlands and thus worldwide' (translated from the Dutch original by the author van Lier 2005, 11).*

The cooperative in Aalsmeer represents the interests of its members (i.e. collective power) as well as its own interests. This actor's political and economic power comes into play through its role as a discussion partner for banks, authorities, and politicians.

5.5 Period IV: Repositioning Within the Chain (1980s to 2000s)

During the transition from period III to period IV, Dutch cut flower growers were under increasing economic pressure. The global oil crisis of the late 1970 s and early 1980 s had an adverse effect on the entire Dutch economy (International Monetary Fund 1999). As early as the 1950 s, 'Dutch suppliers began serving foreign locations, and there was dissemination of production technology, information on best practices, and product innovations. Dutch companies invested in foreign production, logistics, and supplier subsidiaries (Porter, Ramirez-Vallejo, and van Eenennaam 2013, 6). However, influenced by the external shock of the energy crisis, the spatial structure of the cut flower commodity chain changed fundamentally (Porter, Ramirez-Vallejo, and van Eenennaam 2013, 6). Fixed production costs, as well as logistics costs, were rising (Gebhardt 2014, 215). The cut flower sector is highly dependent on production processes such as heating and cooling, as well as the manufacture of fertilisers and pesticides, and the transport of oil (Gebhardt 2014, 100). A lack of market acceptance for higher prices means value creation is declining, particularly for farmers (economic power), and the overall market situation is characterised by uncertainty. In period IV, the focus of Dutch actors shifted towards research, development and trade—areas in which capital accumulation is higher.

Conversely, in the late 1970, European cut flower producers began to relocate their production to regions with more favourable conditions. Initially, mainly to Kenya and Ecuador (Former Manager 2022). These regions were also attractive for production due to low labour costs and minimal government regulation (Gemählich 2022, 87–88). The consistently good climatic conditions throughout the year resulted in lower energy costs for operating greenhouses, which are the largest cost factor for Dutch production. However, there was also the additional cost of air transport in terms of both costs and uncertainties in handling. In the medium term, both processes improved the economic power of flower growers by enabling them to influence production costs more effectively.

In period IV, an internal process can be observed in the growth of market share for production in peripheral regions of the modern world-system. The first cut flower production in Kenya began in the 1980s on Lake Naivasha, on a farm founded by a Dutchman (Hayes 1997, 350). Between 1980 and 1990, cut flower production in this region grew rapidly, increasing from 429 hectares in 1982 (Gemählich 2022, 87) to 4,450 hectares in 1990 (Mekonnen, Hoekstra, and Becht 2012, 3727). However, transporting these fragile goods initially posed a challenge. For example, the first harvest from Osarian Farm (the first cut flower farm at Lake Naivasha, Kenya) was rejected in the Netherlands due to poor quality: 'It had been delayed so long on the journey from Naivasha that the flowers were almost composted' (Hayes 1997, 350). In response, the farm set up its own refrigerated facility with offices to handle freight at Nairobi Airport, and later its own logistics firm (freight forwarder (focused KEN to NL) 2022b; 2022a). Refrigerated trucks were also purchased to transport goods between the farms in Naivasha and the airport (Hayes 1997, 360).

A controversial issue at the flower auction in Aalsmeer during period IV was the admission of imported goods. Non-European imports 'were viewed as a threat, despite accounting for no more than 8% of total auction sales in 1979' (van Lier 2005, 40). Import restrictions were introduced in 1994 and approved in spring 1996. Initially, however, producers from non-European countries were unable to sell their goods at Dutch flower auctions. This was probably made possible by two processes. First, European/Dutch producers were struggling to meet the growing demand, particularly during the European winter. Products from non-European countries were therefore intended to stabilise the volume of goods on the market throughout the year, thereby stabilising the market itself (Gebhardt 2014, 215). Second, non-European producers sought alternative distribution channels due to being denied market access via the auction. This led to the development of the platform Tele Flower Auction in 1995 (Gebhardt 2014, 101). This digital platform was growing into a major competitor to the flower auction in Aalsmeer. The cooperative VBA was able to stabilise and expand its political power over market access and the regulation of the market and transactions by admitting non-European producers in 2006 (Gebhardt 2014, 100). Alongside international flower producers, international buyers also flocked to the auction, further strengthening Bloemenveiling Aalsmeer's position as an international marketplace (van Lier 2005, 40; Porter, Ramirez-Vallejo, and van Eenennaam 2013, 7; Gebhardt 2014, 100; Gemählich 2022, 59).

Furthermore, FloraHolland purchased the Tele Flower Auction in 2010, which no longer competes with FloraHolland's marketplace. Overall, the share of 'internationally produced horticulture remained under thirty percent of turnover until the new millennium' (Gebhardt 2014, 100). During period IV, the commodity chain expanded, partly due to the geographical relocation of production processes. This enables the cooperative flower auction in Aalsmeer to strengthen its position within the commodity chain by incorporating non-European members. This spatial expansion created a link between the core region of the Netherlands, a former colonial power, and peripheral regions such as Kenya (formerly British) and Ecuador (formerly Spanish), which previously had no direct relationship with each other. Conversely, some actors lost political power in the form of influence over and regulation of production processes as a result of the relocation of production processes, including state organisations, unions, and Dutch employer and industry associations.

The EU gained political power in the form of regulation, which gave it influence over non-European production through the imposition of import regulations. As a result, state control organisations were established along the entire commodity chain, primarily to ensure compliance with EU regulations (Lead Auditor 2022; Chief Inspector 2022; Responsible Official 2022). Furthermore, state actors repeatedly influenced the power of airlines and their position in the cut flower commodity chain. On the one hand, state-owned airlines were subsidised. On the other hand, prices for customers were reduced significantly, cargo terminals were expanded (e.g. Frankfurt) and landing fees were increased or decreased for individual airlines. Despite various disputes, however, the 'Koninklijke Luchtvaart Maatschappij' (KLM) remains the main airline and Schiphol Airport the main import and export hub for the cut flower commodity chain (Gebhardt 2014).

Additionally, state actors in the periphery influence the spatial structure of the cut flower commodity chain by promoting export-oriented agriculture through development strategies (Gemählich 2022, 87–88). At the same time, these actors' passivity with regard to regulation in the establishment of production sites is described as beneficial.

Dutch farmers have adapted to these internal and external processes by changing their approach to cut flower production. On the one hand, production shifted towards high-value varieties, such as the Red Naomi rose (Auctioneer 2022). On the other hand, they are focusing on the trade and distribution of imported goods, as well as the research and development of new varieties and cultivation techniques. Evidence of this can be seen in the production value of Dutch flower producers. Between 2003 and 2007, it increased by 10% each year, while greenhouse production decreased by 3% each year (Porter, Ramirez-Vallejo, and van Eenennaam 2013, 7).

5.6 Period V: Standardisation (1990s Until Today)

The transition from period IV to period V is characterised by several external processes. On the one hand, certificates were introduced as a closure mechanism within the structure of actors (political power). This was partly a reaction to growing awareness among end consumers (external lobbying—political power). On the other hand, food retailers entered the cut flower commodity chain.

The Dutch flower industry founded MPS in 1993. This organisation certifies cut flower producers and was initially intended to protect the Dutch flower production. It was designed to shield domestic producers from foreign competition. It was assumed that non-European producers would require time to meet the necessary standards. However, the opposite was the case: producers in Kenya, for example, were able to adapt their processes quickly and thus obtain the necessary certification in a short period of time (Former Manager 2022). 'Nowadays, [MPS] is the minimum requirement for auction sales and the only label shown on the clock' (Gemählich 2022, 77). Demonstrating the political power of MPs within the cut flower commodity chain. Other certificates, such as FairTrade-certification, have emerged in response to increased consumer awareness of production conditions in countries such as Kenya. This awareness has developed through media coverage (Porter, Ramirez-Vallejo, and van Eenennaam 2013: 14–15), Research work, but also documentation, such as 'a Blooming Business' (van Zandvoort 2009). In addition, the Swiss Max Havelaar Foundation (Fair Trade) has launched campaigns to raise consumer awareness in Europe about production conditions. The FairTrade-label originated as the 'Blume' project in 2000 (Former Manager 2022). In addition to these certifiers operating internationally, some operate at a state level. For example, the Kenyan Flower Council

(KFC). The KFC was founded in 1996 to represent Kenyan cut flower growers (political power) (Kenya Flower Council, n.d.; Lead Auditor 2022). With their certificates—KFC silver and KFC gold—the KFC 'adopted codes of conduct and compliance with environmental standards and labor certifications' of MPS standards (Porter, Ramirez-Vallejo, and van Eenennaam 2013, 14–15).

The entry of food retailers into the cut flower commodity chain led to significant changes in the structure of actors and inherent power structure during period V, influenced by the certification movement. Additionally, product characteristics are evolving (Gemählich 2022, 8). Kenyan cut flowers were sold in British supermarkets for the first time in the 1980 s (Gemählich 2022, 5): 'the tendencies are identical all across the EU: supermarkets increase their share, while florists' sales drop' (Gemählich 2022, 56). Cut flowers are sold in food retail outlets in a completely different way to how they are sold in specialist shops. The latter are the traditional distribution channel and the final link in the chain before the end consumer:

> *'usually offer a broad range of flowers, including specialties, new and innovative varieties, and expensive high-quality flowers. In comparison, the retailers in the unspecialised market channel usually sell a narrow product range and highly standardised small to medium sized flowers. This specialisation and segmentation entailed a new valuation of flowers and resulted in changes to breeding, handling, branding, and a new delimitation of the item for sale from single stems to bouquets' (Gemählich 2022, 75).*

Due to food retailers' growing market share, there is an increasing demand for uniform cut flowers that are identical in shape and size (Gemählich 2022, 75). Similar to other agricultural products in the fresh produce segment (like apples see f.e. Hering 2022, 133 ff).

At the same time, the shift towards digital auctioning in Aalsmeer means that traders must form an impression of the commodity based on photos. In some cases, photos from databases are used instead of images taken on the day (Upper Management 2022c; Logistic Expert Royal FloraHolland and Docservice Manager 2022). The changing requirements for product properties are impacting the cut flower commodity chain. Special varieties of high-end roses, primarily produced in the Netherlands, are in increasingly lower demand. Cut flowers, especially roses, became mass-produced in period V: 'global flower exports has leaped from one billion to US $ 8.5 billion within 30 years between 1988 and 2020' (Gemählich 2022, 5). However, other cut flowers, such as tulips, can also be cultivated on an industrialised and standardised scale in modern greenhouses, regardless of external conditions (Upper Management 2022b). Additionally, the nature of the relationship between producers and intermediaries is changing due to the preference for long-term direct trade expressed by food retailers and certification groups.

The effects of the external shock caused by the global economic crisis of 2008 on the cut flower commodity chain can be inferred. While the sales market grew little between 2007 and 2009, it then rose quite sharply (Art Blossom 2020).

The cut flower commodity chain has also changed during period V due to increasing digitalisation. The Aalsmeer flower auction, which traditionally served as a physical marketplace for presenting and pricing flowers, has been conducted exclusively online since 2021 (Auctioneer 2022). Figure 5.5 shows the virtual auction, and Fig. 5.6 a former auction hall. Both are located in Aalsmeer.

While personal inspection of the goods used to be central to the purchase process, standardisation and certification of the goods, as well as personal networks and experience, are now of crucial importance. Additionally, Royal FloraHolland has operated the digital platform Floriday for direct trade since 2018 (Royal FloraHolland. n.d.). Despite the marketplace shifting to the digital realm, the town Aalsmeer remains at the centre of the cut flower commodity chain.

In spring 2022, two years after the start of the external shock of the COVID-19 pandemic, cut flower producers in the periphery are considering ways to diversify their businesses. This applies to both their distribution channels (via the auction and direct sales) and their transport routes (air and sea). This is intended to secure their position in the event of external shocks. Furthermore, it will reduce their costs and ensure they continue to create value. During COVID-19 the food retail has strengthened their economic power. While the flower auction in Aalsmeer and flower shops worldwide were temporarily closed, direct contracts were unilaterally cancelled by food retailers. During the first lockdown, flowers were destroyed at auction many points of the commodity chain. Consumers only left their homes for essentials, changing their shopping behaviour. Additionally, transport options for cut flowers decreased due to the decline in passenger flights (Espeter and Retamal 2024).

In addition, production and transport costs are rising more sharply as a result of the trade embargoes imposed in response to Russia's war of aggression, particularly with regard to fertilisers, pesticides and energy prices (Managing Director 2022). Furthermore, such external shocks are affecting access to sales markets. During the COVID-19 pandemic, there were hardly any passenger planes, which were the main means of transporting cut flowers globally before the external shock. Scarce transport capacities mean that, since the Russian invasion of Ukraine, cut flowers are being imported via other airports. However, due to their geographical location, the producing regions are closely linked to their target

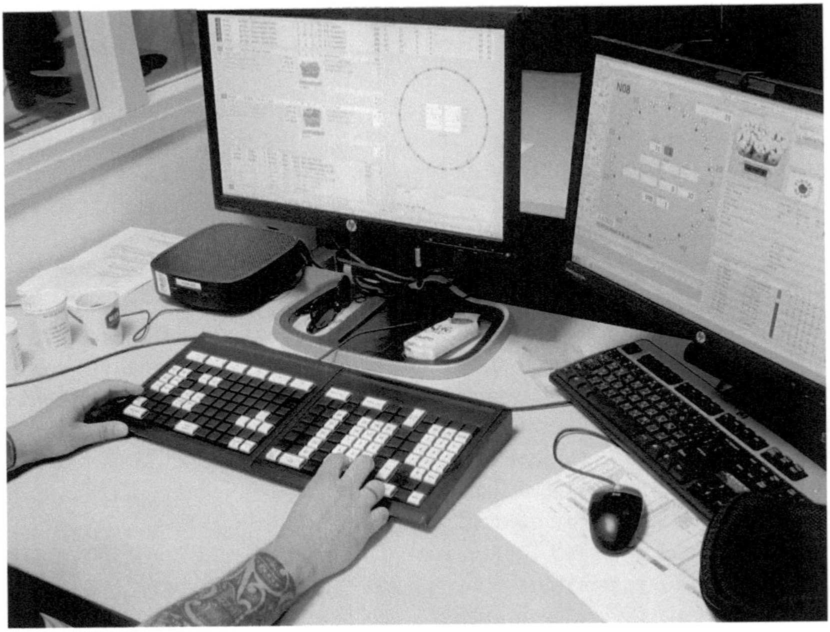

Fig. 5.5 Auctioneer during a virtual auction in Aalsmeer (Engel 2022)

markets. Their production is geared towards these markets. Dependence on the stability of the relationship between production and trade has increased enormously, especially for flower producers in period V (reduction in political and economic power).

Fig. 5.6 Former auction hall in Aalsmeer (Engel 2022)

The Cut Flower Commodity Chain in the Longue Durée 6

'In less than a century, this little town with a cottage flower industry mushroomed to become the high-tech, powerful center of global horticultural commerce and innovation. Aalsmeer was quite a different place before the auction, and before residents began in earnest to cultivate roses and other flowers' (Gebhardt 2014, 89).

In the previous Chap. 5, I examined the process-nature of the economic structure cut flower commodity chain. My focus was on the external and internal processes within and along meaningful time periods. This enabled me to demonstrate the influence of the processes involved in establishing and maintaining power within the structure of the cut flower commodity chain. Regardless of these processes, it became clear that the flower auction in Aalsmeer was able to establish, expand and maintain its position within the inherent power structure during the time under study.

In this chapter, my analysis of the cut flower commodity chain focusses on the reconstruction of the period-specific situation within the meaningful periods identified in the periodisation. While the examination of the formative phase focuses on the entire region of today's Netherlands, the analysis of the individual periods places particular emphasis on the Aalsmeer region. Each period is examined by first describing the historical context, followed by the analytical step of reconstructing the structures of actors, spatial structure, and inherent power structure.

To improve understanding of the text, I will discuss this analytical step second.

L. M. Espeter, *From Hobby to Billion-Dollar Industry*, Wirtschaft + Gesellschaft,
https://doi.org/10.1007/978-3-658-52437-1_6

6.1 Period I: Start of Commercial Cut Flower Production (1850s to 1910s)

The first period is characterised by the emergence of the commercial cut flower production. This was driven by the expansion of agricultural land and the growing market for cut flowers.

6.1.1 Historical Context

The Netherlands gained independence from Belgium in 1839 with the signing of the Treaty of London. From then on, it became known as the Kingdom of the Netherlands. Between the 1850s and the 1910s, the Netherlands was characterised as a state by economic growth and the expansion of its colonial territories (Kennedy 2017; Wintle 2000). The Netherlands has generally been one of the richest countries in the world since the 19th century, with wealth concentrated in urban centres and supported by the surrounding countryside (Wintle 2000, 101).

Unlike Great Britain, for example, the Netherlands did not experience a comprehensive Industrial Revolution. Consequently, the manufacturing sector is relatively small. Instead, sectors such as agriculture, trade, and finance have dominated the Dutch economy (Wintle 2000, 75–77). Analysis of the economic structure of the Netherlands shows that the proportion of people employed in agriculture and fishing fell from 43.8% in 1849 to 28.3% in 1909 (Wintle 2000, 78). During this period, the proportion of people employed in trade remained relatively constant at around 22% (Wintle 2000, 186).

Among other things, economic growth is reflected in significant population growth. For example, between 1849 and 1909, the population nearly doubled from around 3.3 million to 5.9 million (Wintle 2000, 9). As the population increased, so did the size of the country's available labour-power.

Regional differences in average agricultural wages are evident. In North Holland, for example, they rose from 80 cents in 1850 to 187 cents in 1910. Throughout the entire period, salaries in North Holland were average compared to the other provinces and, from 1880 onwards, higher than in all other provinces in the country (van Zanden 1987, 117).

In the second half of the 19th century, the state made significant investments in infrastructure, particularly transport networks and canal construction, with a focus on connections to Germany (Wintle 2000, 95). In the mid-19th century, favourable weather conditions and the introduction of artificial fertilisers and other agricultural innovations boosted the agricultural sector (Wintle 2000, 54). Overall, the

Netherlands experienced considerable economic growth in the years leading up to the First World War (1914–1918) (Kennedy 2017, 327).

6.1.2 Structure of Actors

At the beginning of the commercial cut flower production, farmers in the Aalsmeer region grew cut flowers as a by-product. Around 1860, there were approximately one hundred flower growers in the this area. Due to the translation, it is not possible to distinguish between growers and producers ('bloementeelt' means both flower cultivation and production), but it is likely that this distinction did not yet exist in period I. The flowers grown were mainly lilies, carnations, lilacs and roses, with many growers specialising in roses (Gebhardt 2014, 89). Initially, these were sold alongside fruit and vegetables at local markets (*Enkhuizer Courant* 1883). However, as large cities grew from the mid-19th century onwards, intermediaries emerged in the form of middlemen and exporters (van Lier 2005, 21–22). By the end of the 19th century, some of the flowers grown were destined for export (*Enkhuizer Courant* 1883) (Fig. 6.1).

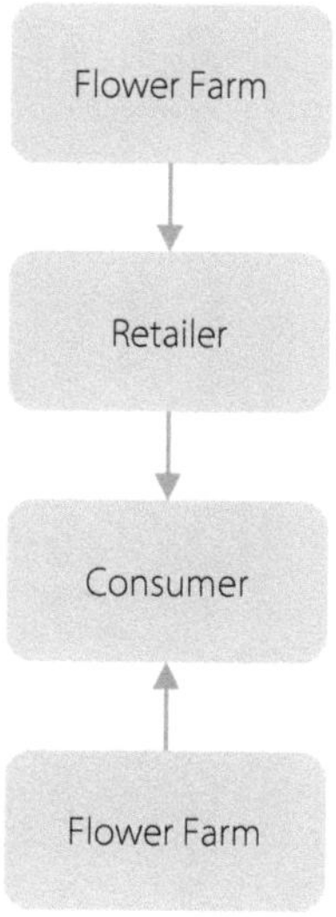

Fig. 6.1 Visualisation of the reduced structure of actors in period I. (Own illustration)

At that time '[m]ost florists were retailers and small growers with two or three greenhouses just outside the urban areas' (Ziegler 2007, 33). Cut flowers were sold as individual stems or as arrangements. By the end of the 19th century, 200 heated greenhouses were being used for cultivating flowers in Aalsmeer (*Schuitemakers Purmerender Courant* 1886). The state secondary school for floriculture also opened in Aalsmeer during this period as the Rijks Middelbare Tuinbouwschool voor Bloementeelt te Aalsmeer (Nationaalarchief 2025b), providing a local centre for education and research in this field ever since.

6.1.3 Spatial Structure

Focusing on the spatial structure of the cut flower commodity chain reveals that historically important cut flower regions differ in terms of their location conditions. The region around Haarlem and Leiden, with its sandy soils, is ideal for bulbous plants. In the Aalsmeer region, however, these were hardly cultivated at all. Instead, greenhouses were built on the drained peatlands around the newly created lakes, where lilies, carnations and roses were mainly grown (Gebhardt 2014, 89; Historische Tuin Aalsmeer 2020).

Fig. 6.2 Transportation of cut flowers via canals in Aalsmeer around 1950. (Source: Noord-Hollands Archief, G.T. Wenzel)

Fig. 6.3 Visualisation of the spatial structure in period I. (Own illustration)

Due to their highly perishable characteristics, cut flowers could only be distributed over a limited area. Farmers harvested the cut flowers in the early hours of the morning so that they could be sold in the surrounding towns that same morning. Most of the cut flowers sold in Amsterdam were produced in Aalsmeer due to its proximity (see Fig. 6.3.) The canals were used to transport flowers on small boats (see Fig. 6.2, and van Heck 2021, 87).

As well as transporting goods, the canals created by drainage projects also played a central role in heating the greenhouses. Until the mid-20th century, these were heated with coal, which was transported via the canals (van Heck 2021, 53). Coal was previously imported from Belgium and Germany. Mining in the Limburg region, which has been taking place since 1900, only covered 7% of the country's needs (Wintle 2000, 109). However, it can be assumed that the peat available in the Aalsmeer region (Wintle 2000, 110) was also used for heating.

6.1.4 Inherent Power Structure

Examining the inherent power structure of the cut flower commodity chain during period I in terms of political and economic power reveals two key findings: there was minimal political influence, and farmers held little influence over market pricing. The power structure in period I was characterised primarily by the economic power of intermediaries. Market pricing was the dominant factor in the cut flower commodity chain. Initially, flower cultivation was only a source of additional income for farmers and gardeners. However, market demand was unstable and competition from other farmers was high (Gebhardt 2014, 89 ff). The steady expansion of cut flower production led to the market becoming partially flooded (Gebhardt 2014, 93). This enabled intermediaries with good connections to the consumer market to push down prices, thus reducing profits for farmers (Gebhardt 2014, 28; 2014, 94).

However, cut flower producers had already exerted considerable influence over political decision-making by the end of the 19th century, when a state secondary school for floriculture opened in Aalsmeer. The choice of this location does not seem arbitrary, given that Aalsmeer was already renowned throughout the country for its flower cultivation at the beginning of the 19th century (*Enkhuizer Courant* 1911; van Lier 2005, 27–28).

> *A Parisian florist who visited Aalsmeer reports on the local flower industry: 'I am thrilled by what I have seen, but I am also ashamed that we French are still 100 years behind in flower cultivation. [...] In Holland, flowers come from huge factories with large chimneys. Roses are produced there with the help of thousands of tonnes of coal. They are manufactured like nails or car tyres. That may not be poetic, but it is modern. In Holland, they are not afraid to spend a million guilders on building greenhouses. And the result is that the flowers are less bland than ours, that you no longer have to worry about changeable weather, that night frosts can no longer ruin the harvest*

and that prices no longer rise and fall with the thermometer. The flower growers in Aalsmeer are industrialists. They are not only organised for the mass production of flowers, but have also perfectly organised the sale and transport of their goods' (R. 1928 - translated from the Dutch original by author).

6.2 Period II: Establishment of Central Locations of Trade (1900s to 1950s)

Period II is characterised by the establishment of central trading centres for cut flowers, which gave the period its name. Between 1911 and 1929, seven flower auctions were set up in the Netherlands, and this number had already risen to seventeen by 1945. The first of these were Bloemenlust and Centrale Aalsmeerse Veiling (CAV), both in Aalsmeer (Gebhardt 2014, 95). This period was also marked by increasing political power along the chain and external shocks. The period was bounded by the establishment of the first flower auctions in the winter of 1911/1912, and the post-war years. In the 1920s, Aalsmeer was considered the centre of flower cultivation (*Deventer Dagblad* 1923; van Lier 2005, 27–28). It continued to consolidate its position during period II.

6.2.1 Historical Context

The first half of the 20th century in the Netherlands was significantly shaped by external shocks, including the First World War (1914–1918), the global economic crisis of the 1930s, and the Second World War (1939–1945).

During this period, the cooperative movement in the Netherlands experienced a significant upswing, with roots in the late 19th century (Gemählich 2022, 58). Faced with hard, poorly paid and unregulated agricultural work, people sought protection from an exploitative system by forming collective organisations (Gebhardt 2014, 28). It was in this environment of agricultural cooperatives, industrialisation, and egalitarianism that fruit and vegetable auctions first emerged, followed later by the flower auctions. The Dutch government supported this development by making the use of cooperative auctions compulsory in the early 20th century. After their initial establishment, agricultural auctions experienced a period of rapid growth. By 1899, around 924 cooperatives had already been registered nationwide. By 1913, this figure had risen to 1,177, with over 100,000 members (Wintle 2000, 177).

The total population of the Netherlands grew steadily (1909: 5,9 Millionen; 1949: 9,6 Millionen; 1960: 11,5 Millionen) (Wintle 2000, 9), which continuously increased the availability of labour-power.

At the same time, the trend of moving away from agriculture to other economic sectors continued. The proportion of people employed in agriculture and fishing fell from 28.3% in 1909 to 19.2% in 1947, dropping further to 10.7% in 1960 (Wintle 2000, 78). The First World War presented the Netherlands with economic and social challenges, as it maintained its neutrality under difficult circumstances and thus avoided invasion (Kennedy 2017, 349 ff). The country was hit by the global economic crisis between 1933 and 1936 (vaan Oostrom 2011, 84). The years 1940 to 1945 were marked by the German occupation of the Netherlands. The German attack on 10 May 1940 and the subsequent bombing of Rotterdam on 14 May resulted in the surrender of the Dutch armed forces and government (Kennedy 2017, 370). The liberation by the Allies began in autumn 1944 and was completed in May 1945 (Kennedy 2017, 380).

Fig. 6.4 Central hall of CAV 1915. (Source: Noord-Hollands Archief, Public Domain)

The Second World War and the subsequent occupation had a significant impact, resulting in the partial collapse of the flower auction industry, for example. Certain types of flowers, such as lilacs, carnations and roses, were excluded from auctions in 1942 due to high production costs (Gebhardt 2014, 104).

All three external socio-political shocks — the First World War, the financial crisis of the 1930s, and the Second World War — resulted in severe food shortages, since the Netherlands had previously prioritised growing lucrative export crops over ensuring its own food supply's (Wintle 2000, 177–178).

6.2.2 Structure of Actors

The establishment of central marketplaces, such as flower auctions, transformed the structure of the industry from period I to period II. Cooperatively organised auctions now act as intermediaries between flower producers and traders. In addition to their distribution method, the cooperative structure is a distinctive

Fig. 6.5 Auction hall of Bloemenlust 1915. (Source: Noord-Hollands Archief, Public Domain)

feature of flower auctions. These marketplaces are owned by the flower producers themselves. As members of the cooperative, they are required to sell their goods through it. In return, they gain the right to vote on the cooperative's decisions.

The first flower auction was founded in Aalsmeer in 1911 as Bloemenlust. A few days later, the CAV was established. There were two flower auctions in this small region for the following reasons: "[the] Bloemenlust growers were Dutch Reformed Protestants and Catholics, while the CAV group were Baptists and socialists, and each circulated in their own social circles, pubs, schools, and sports clubs" (Gebhardt 2014, 97). Kennedy writes: 'to the outside world, Catholics formed a single and closed front' (Kennedy 2017, 341). At the time, this division between religious communities was also reflected in associations and political parties (Kennedy 2017).

Figure 6.4 shows the central hall of the flower auction CAV in 1915, where cut flowers were traded before and after the auction. While Fig. 6.5 shows the auction hall of the flower auction Bloemenlust in 1915.

Furthermore, firms specialising in specific parts, such as breeding, are emerging along the commodity chain (Gebhardt 2014, 90 ff). Kweekster are first mentioned in newspaper articles. In addition to the state secondary school for floriculture, which was founded in 1897, the 'Tuinbouwinternaat voor Meisjes' (horticultural boarding school for girls) was established by 1951 to provide a three-year training course in plant and flower cultivation following secondary school (*Utrechtsch Nieuwsblad* 1951).

In addition, supraregional interest groups existed in the cut flower industry during period II. Examples include the Katholieke Nederlandse Boeren en Tuindersbond (Catholic Dutch Farmers' and Gardeners' Union) and the Nederlandse r.k. Bloemistenvakbond (Dutch r.c. Florists' Union) (Fig. 6.6).

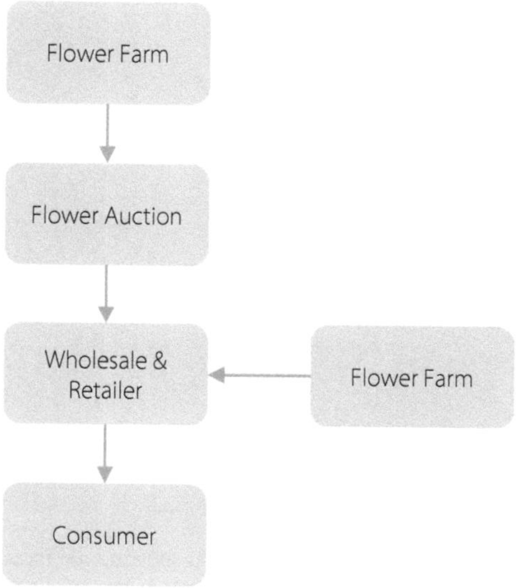

Fig. 6.6 Visualisation of the reduced structure of actors in period II. (Own illustration)

6.2.3 Spatial Structure

The spatial structure also reflects the separation between CAV and Bloemenlust. I already discussed the social aspect. In terms of geography, the flower auctions are located in different parts of Aalsmeer. CAV was located on Marktstraat, later on Van Cleeffkade, and Bloemenlust was located on Oosteinderweg. The distance between them is just under 3 km. The Jaap Bollebrug (English: bridge) lies between them, marking the geographical boundary between the social groups in Aalsmeer at that time into Aalsmeer 'durp' and Aalsmeer 'farregat' (van Lier 2005, 31). These locations are illustrated in Fig. 6.7. It can be assumed that the two auctions merely continued the existing spatial and geographical separation within Aalsmeer.

The addition of flower auctions did not significantly alter the spatial structure. Due to the highly perishable nature of the goods, gardeners and farmers set up the auctions in their immediate locality. However, the process of transferring goods from producer to trader changed. While in period I producers from Aalsmeer brought their goods to Amsterdam themselves, traders now had to come to Aalsmeer, bid for the goods there, and transport them to the city themselves. Initially, the auctions were held in cafés or pubs, where the flower producers

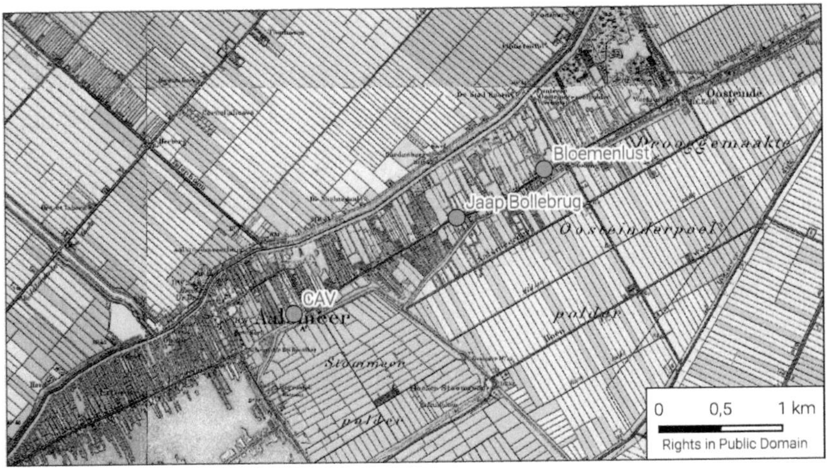

Fig. 6.7 Map of Aalsmeer 1900 showing the location of the CAV, the Jaap Bollebrug, and the Bloemenlust. (Own illustration, source of the map: Nationaal Georegister)

would spread their goods out on tables, sorted by type, and offer them for auction (burocontent 2021; 'Largest Flower Auction', n.d.). The aerial photographs from the late 1920s illustrates the spatial integration of CAV and Bloemenlust (Fig. 6.8 and 6.9).

In addition to their geographical proximity to the production sites, the auction buildings were also 'located near waterways, allowing flowers to be delivered locally by boat and bicycle and serving as gathering points for the production of surrounding growers. Railways provided efficient transportation between the auctions and the Amsterdam transportation hub' (Porter, Ramirez-Vallejo, and van Eenennaam 2013, 6). The Figs. 6.8 and 6.9 show the proximity of the flower auctions to the canal network in the end of the 1920s. The canals served as the main transport route for producers and traders during period II (*Maas- En Roerbode* 1947).

In addition, the 'Speciale Bloementrein' (engl.: Special Flower Train) is used within the Netherlands, while the aircraft is used for exports (*Maas- En Roerbode* 1947). Amsterdam's Schiphol Airport began operating in 1916 as a military base, previously known as Fort Schiphol ('Kaart van Den Ringdijk Tusschen Schiphol En Het Nieuwe-Meer, Mijlpalen 35–37 Waarop Met Kleuren Zijn' 1867; Taurel. 1867). Schiphol was also used for civil aviation from 1920 onwards (*Het Nieuws* 1920). Schiphol Airport is located on the site of the former Haarlemmermeer

Fig. 6.8 Aerial photograph of flower auction CAV around 1925/1930. (Source: Noord-Hollands Archief, KLM Aerocarto)

(*Algemeene Weekblad* 1857). The flower auction in Aalsmeer is located close to Amsterdam Schiphol Airport, which made it easier to start exporting to countries such as England and Germany (Gebhardt 2014, 97, see Fig, 6.10). Founded in 1919, KLM was already flying daily to London by 1920 (Wintle 2000, 203). Cut flowers from Aalsmeer have been exported to London since 1922 (Royal FloraHolland. n.d.).

In 1947, exporters travelled to meet their foreign business partners, taking with them colour photographs of the available cut flowers, which the partners used to inform their purchasing decisions.

> *'Aalsmeer also makes a significant contribution to foreign trade, and large quantities of flowers are exported to England, Belgium and Scandinavia, especially in winter. In one department, we saw girls packing lilies that would be on display in shop windows on Fifth Avenue in New York 24 hours later. Even though KLM's flower express to America is still more of a symbolic gesture, the name Aalsmeer floats around the*

Fig. 6.9 Aerial photograph of flower auction Bloemenlust 1926. (Source: Noord-Hollands Archief, Public Domain)

> *world in the hope that Uncle Sam will one day really take off' (*Maas-En Roerbode *1947 – translated from Dutch original by author).*

In 1949, wholesalers and retailers from across the country bought flowers at the auctions in Aalsmeer: 'Exporters act on behalf of their business partners in other European countries. Recently, large quantities of flowers have also been exported to the United States [transport via Schiphol with KLM]' (*De Zaanlander* 1949—translated from Dutch original by author) (Fig. 6.11).

The cultivation and breeding of flowers in Aalsmeer now covered in the 1940s over 13 km2 of greenhouses and flowerbeds. According to a newspaper report, this makes it the largest glass-covered cultivation area in Europe (*De Zaanlander* 1949). Figure 6.12 shows a photograph of the greenhouses in Aalsmeer from that period.

Fig. 6.10 Export via Bloemenexpress and KLM 1928. (Source: Noord-Hollands Archief, N. Joore)

6.2.4 Inherent Power Structure

For the first time, political power played a central role in the inherent power structure of the cut flower commodity chain in period II. The industry's growing importance to the Dutch economy is reflected in the increasing number of political decisions made in its favour during this period. Therefore, it can be assumed that actors such as the cooperative flower auctions, flower growers, breeders and traders wielded political power in the form of political influence. At the same time, this political influence can be traced back to the historical links between the flower industry and the country's politics and finance (see Sect. 5.1)

In period II, political power in the form of market access regulation was exercised by the cooperative flower auction. Cut flower producers must be members of the cooperative to sell via their marketplace. They are also obliged to offer their goods exclusively at the auction. In this way, the cooperatives regulate who can sell cut flowers on the market, while also centralising it to specific marketplaces.

Although there were other cooperative flower auctions besides the two in Aalsmeer,'Rijnsburg (west) in 1914, in Blerick (south) in 1915, in Poeldijk (west)

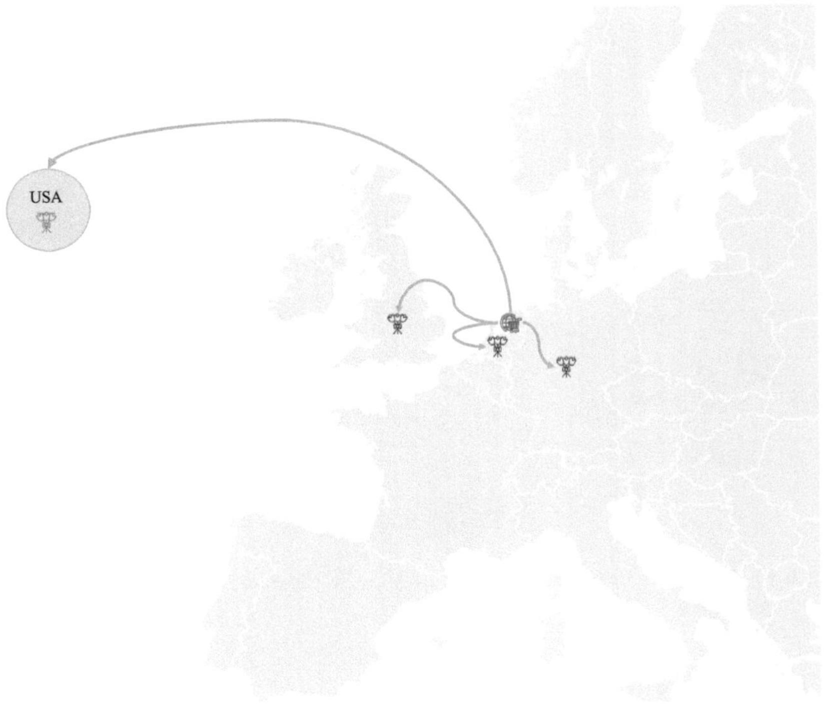

Fig. 6.11 Visualisation of the spatial structure in period II. (Own illustration)

in 1923, in Eelde (north) in 1927, and in Berkel en Rodenrijs, near Rotterdam (west) in 1929' (Van Heck 2021, 91) (Fig. 6.13).

'But from the beginning, in many ways Aalsmeer led the industry. [...] [I]t had not only the largest market share, but also excelled with the best established export position' (Gebhardt 2014, 95–96).

The dominant position of the Aalsmeer auctions is evident from the turnover figures for the first financial years: 'Bloemenlust and the CAV enjoyed tremendous success in both domestic and international trade, earning in their first year 110,000 and 80,000 guilders' (Gebhardt 2014: 96). 'In 1918, the Central Aalsmeer Auction achieved—for the first time—annual sales of one million guilders (the equivalent of €6.8 million nowadays)' (van Heck 2021, 91).

The establishment of flower auction cooperatives has significantly altered the inherent power structure compared to the first period. Economic power, in the

Fig. 6.12 Greenhouses for cut flower production in Aalsmeer around 1940. (Source: Noord-Hollands Archief, Public Domain)

form of influence over pricing, has shifted from traders to flower growers. Pricing through the Dutch auction system enabled quick sales at market prices. The flower auctions stabilised market prices and 'strengthen the position of growers' (Gebhardt 2014, 95). In 1929, there were seven flower auctions in the Netherlands. By 1945, this increased to 17 (Meulenberg 1989). It can therefore be assumed that the power of individual flower auctions is declining. They compete with each other, and traders can choose which marketplace to use.

The significance of the Aalsmeer region for cut flower production is evident from the fact that, during the global economic crisis of 1933–36, subsidies were granted to producers in Aalsmeer and Haarlem, but not in other regions. This suggests that producers in Aalsmeer had good connections to political actors. Both signs of political power.

At the same time, reports from the post-war period state that 'while knowledgeable growers placed their village by far at the forefront of Dutch flower

Fig. 6.13 Locations of flower auctions 1933. (Own illustration)

producers. Aalsmeer supplies more than half of the total Dutch flower productic per year' (Maas-En Roerbode 1947—translated from the Dutch original by the author). Therefore, it can be assumed that flower producers in Aalsmeer are strengthening their dominant position in the cut flower commodity chain. This is partly due to their market dominance, but also because they are renowned for the quality of their products. However, during the post-war period, cut flower producers were dependent on allocations from relevant government agencies due to a general shortage of coal. In 1947, the production capacities in Aalsmeer could not be fully utilised due to a lack of coal, meaning that the greenhouses could not be heated in winter. This also impacted the flower auctions in Aalsmeer, where hardly any flowers were traded at the time (*Maas- En Roerbode* 1947). Consequently, neither the flower auction in Aalsmeer nor the flower producers were able to exploit their political influence for their own benefit in the immediate post-war period.

The prominence of the flower auctions in Aalsmeer, both within the Netherlands and in terms of the country's international image, is evident from the numerous newspapers reports from the 1930s to the 1950s that mention political and economic delegations from other countries being shown around the Aalsmeer flower auction. Churchill, for example, visited in 1946. Many other tour groups and school classes also visited the flower auction (*Nieuw Noordhollandsch Dagblad* 1946).

6.3 Period III: Economic Growth and Centralisation (1950s to 2000s)

Period III of the cut flower commodity chain is characterised by the merger and subsequent monopolisation. The first merger occurred in 1972, when the two flower auctions in Aalsmeer merged. By 1985, 12 flower auction cooperatives remained in the Netherlands. The final merger occurred in 2008 with the formation of Plantion. Since then, there have been two auction cooperatives in the cut flower sector in the Netherlands: Plantation and FloraHolland.

6.3.1 Historical Context

The period immediately following the Second World War was characterised by stable economic and political conditions in the Netherlands. This resulted in sustained economic growth until the oil crisis of 1973 (Kennedy 2017, 388; 2017, 406 ff).

A pivotal moment for the Dutch economy was the discovery and subsequent development of the Slochteren gas field near Groningen in 1959 (Wintle 2000, 108). This resulted in a shift towards oil and gas as the primary energy sources in the 1960s (Wintle 2000, 114). Revenues from the Slochteren gas fields contributed significantly to financing the national treasury (Kennedy 2017, 407). During period III, the infrastructure was also modernised, for example through the expansion of Schiphol Airport (Kennedy 2017, 400).

In 1960, the total population of the Netherlands was 11,462,000 (Wintle 2000, 9). Based on population figures from 1947 and the 1990s, it is estimated that around 20,000 people lived in Aalsmeer during the 1980s ('Statistieken gemeente Aalsmeer. Aantal inwoners per jaar' 2016). In 1961, the Dutch Reformed Church and the Reformed Churches in the Netherlands called for church unity, leading to greater understanding between these groups (Kennedy 2017, 410).

6.3.2 Structure of Actors

Actors in Aalsmeer continues to play a central role in the cut flower commodity chain. This is a result of the merger that was decided in 1964 and implemented in 1972 as *Cooperative Society Verenigde Bloemenveilingen Aalsmeer* (VBA). Since then, only one flower auction has taken place in Aalsmeer. In the 1980s, over 5,000 growers sold their products via the auction (Patel-Campillo 2010, 90). With the exception of the centralisation of auction cooperatives, the basic structure of

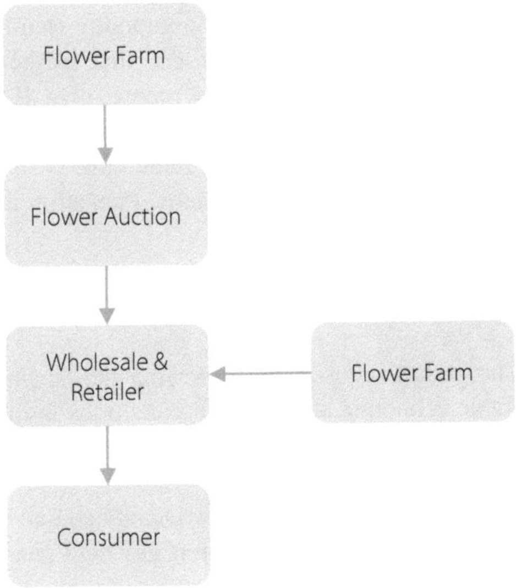

Fig. 6.14 Visualisation of the reduced structure of actors in Period III. (Own illustration)

the market participants remains unchanged between periods II and III. A detailed analysis of the merger dynamics can be found in Sect. 5.4 (Fig. 6.14).

Fig. 6.15 Auction hall of Bloemenlust around 1960. (Source: Noord-Hollands Archief, G.T. Wenzel)

6.3.3 Spatial Structure

The spatial structure of the cut flower commodity chain continued to expand during period III. This was partly due to the centralisation of the industry in

Fig. 6.16 Auction hall of VBA around 1980. (Source: Noord-Hollands Archief, Municipality Aalsmeer)

Aalsmeer, which attracted producers and traders from increasingly distant areas, and partly due to the expansion of the sales market.

In Aalsmeer, the new, much larger VBA building opened in 1972 and is now the central marketplace for the cut flower industry. Figures 6.15–6.18 show the spatial differences between the flower auctions Bloemenlust and VBA, with a time difference of around 20 years. At 303,282 square metres, the auction building of VBA was the largest in the world when it opened in 1982. At that time, the VBA employed around 4,000 people. (*Nunspeets Nieuws- En Advertentieblad* 1982). An increasing number of export and logistics companies were locating around this new building.

In addition, transport infrastructure is being developed to support the flower auction. The roads are being upgraded to cope with the high volume of traffic. The traffic lights around the auction are timed to allow lorries to reach and leave the auction as quickly as possible during peak times.

Fig. 6.17 Inside VBA around 1980. (Source: Noord-Hollands Archief, Municipality Aalsmeer)

> *'Aalsmeer's auctions have been integral to regional planning over the past century. When building highways, intersections, traffic lights and so on, planners have had to take into account the route and schedule of thousands of trucks coming from and going to the auction. Such volumes of traffic also affect quality of life issues for local residents' (Gebhardt 2014, 97).*

Although the European rail network was well developed by the 1960s, growing sales markets in Europe and beyond were still supplied by truck and plane (Porter, Ramirez-Vallejo, and van Eenennaam 2013, 6). This was because transporting goods and passengers on the same train was prohibited. Consequently, in 1982, cut flowers were delivered to the auction by both the producers themselves and a collection service provided by the VBA (*Nunspeets Nieuws- En Advertentieblad* 1982).

Following the merger that led to the formation of VBA, the sales market continued to expand, both in terms of products sold and geographical reach. West Germany remained the main customer. However, export to other countries, such

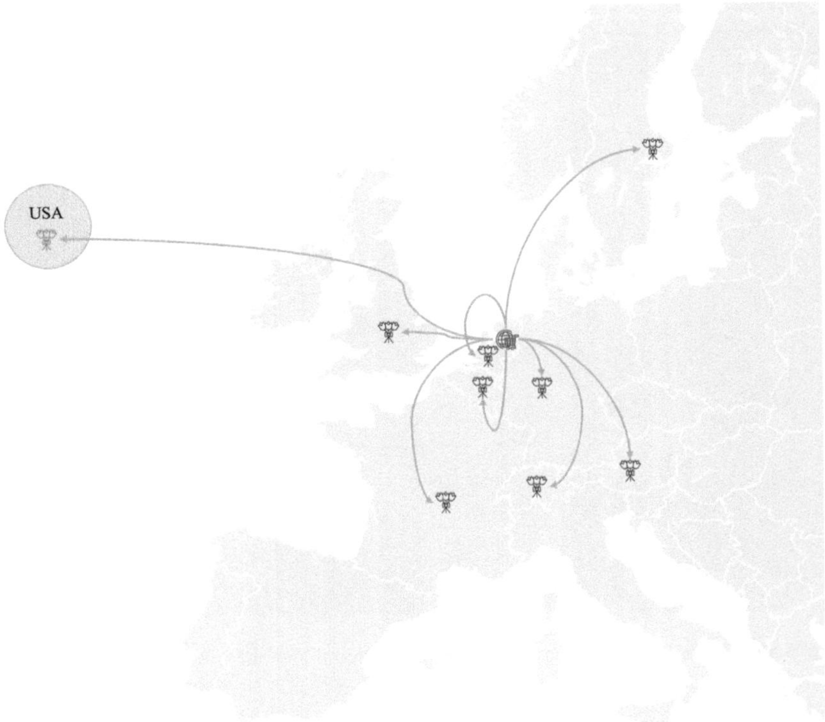

Fig. 6.18 Visualisation of the spatial structure in period III. (Own illustration)

as France, increased (*Betuws(Ch) Tuinbouwblad* 1965a). By 1973, the Netherlands accounted for 71% of global cut flower exports, with exports of flowers and plants worth over 700 million guilders. Italy, the second most important export country, lagged far behind with 14%. The Netherlands mainly exported to Germany (82% of cut flower exports). Other destinations included France, Switzerland, Scandinavia, Austria, Belgium, England and North America. In less than a decade, global exports of cut flowers increased by 175%. By the 1990s, the value of global exports had tripled to reach five billion euros. The Netherlands remained in the lead with a 60 per cent share. The most important markets for Dutch exports in the 1990s were Germany, the UK, France and Italy (van Lier 2005, 24).

6.3.4 Inherent Power Structure

The cut flower commodity chain in period III is characterised by a complex inherent power structures that are evident at political and economic levels. Through several mergers, the cooperative flower auctions have maintained and consolidated their dominant position. For example, the first merger became necessary in the 1960s, as nearby flower auctions in Beverwijk and Vleuten were increasingly competing with those in Aalsmeer (Gebhardt 2014, 97). Bloemenlust's merger with CAV (1964/1974), followed by VBA's merger with FloraHolland (2008/2010), saw the two largest flower auctions in the Netherlands join forces. Steen (2010, 116) argue that growing competition from international flower markets, such as those in Mumbai and Dubai, was another reason for the mergers. As a result of the expansion of its cooperative and the centralisation of the marketplace, the auction's sphere of power grew accordingly. Despite concerns about the creation of a monopoly, the merger of VBA and FloraHolland was approved by the Dutch antitrust authorities in 2010, highlighting the political clout of those involved.

In 2005, the flower auction in Aalsmeer offered commodities from more than 6,000 farmers across fifty countries. Consequently, the flower auction in Aalsmeer dominated the commodity chain (van Lier 2005, 24). The cooperatives themselves determine which goods can be sold on their marketplaces through membership and standard setting. In addition, their members are obliged to sell their goods exclusively there during period III. By regulating and controlling market access, the flower auction in Aalsmeer wields considerable political power, which has an impact on the economic power of those involved. Prices resulting from auctions via the clock in Aalsmeer influence the entire cut flower commodity chain. In addition, the auction cooperative itself earns a percentage of all transactions. Thanks to rapidly growing sales figures, the cooperative's financial resources (economic power) are steadily increasing.

Furthermore, period III was also characterised by increasing diversification among buyers. While exports were heavily focused on West Germany in the 1960s, firms recognised the need to access new consumer markets (*Betuws(Ch) Tuinbouwblad* 1965b). This transition indicates a shift in the political and economic balance of power within the cut flower commodity chain. There has been a decline in dependence on individual sales markets and production regions.

6.4 Period IV: Repositioning Within the Chain (1980s to 2000s)

Period IV is characterised by a relocation of production processes from the European core to non-European peripheries. This is in addition to the realignment of European cut flower producers within the commodity chain.

6.4.1 Historical Context

The energy crisis of the 1970s and the associated massive rise in oil prices impacted period IV. The cut flower industry relies heavily on oil and gas for production, storage, and transport (Gebhardt 2014: 100). Although the Netherlands has its own natural gas reserves, the oil crisis nevertheless led to a slowdown in the economy and higher unemployment (Kennedy 2017: xix, 388/389). In response, the Wassenaar Agreement was adopted to secure the Netherlands' competitiveness through measures such as reducing working hours (Stein 2004).

The end of the 20th century is characterised by the expansion of the EU and the collapse of the Iron Curtain (Britannica 2025). Furthermore, the European Economic Union expanded to become a monetary union in 1999/2000 with the introduction of the Euro (European Union. n.d.-a), results in an even closer integration of the European markets.

The Netherlands was only briefly affected by the global financial crisis of 2008/2009 (Masselink and Noord 2009). The Netherlands' GDP grew from USD 1.377 billion in 1980 to USD 2.636 billion in 2000 ('IMF 2025). In period IV, the number of people employed in agriculture in the Netherlands declined by almost 12%. In 1980, over 260,000 people were employed in Dutch agriculture, compared to just under 230,000 in 2005. The population of Aalsmeer increased from 20,435 in 1980 to 22,662 by 2000 (Netherlands Atlas. n.d.).

6.4.2 Structure of Actors

Compared to period III, the structure of actors changed in that non-European producers became a major part of the cut flower market. They partly replaced the Dutch production process. Nevertheless, the flower auction in Aalsmeer continues to play a central role in linking production and trade (Fig. 6.19).

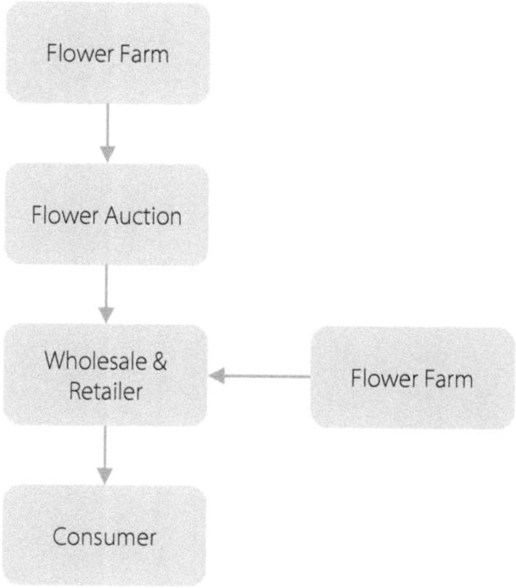

Fig. 6.19 Visualisation of the reduced structure of actors in period IV. (Own illustration)

6.4.3 Spatial Structure

The spatial structure of the cut flower commodity chain is being significantly changed by non-European production. Regions that were not previously involved are now being incorporated into the structure. However, this has merely shifted cut flower production to regions with lower production costs. In the process, new labour and natural resources are being exploited. Input production, trade and sales markets remain spatially unchanged. In period IV, few or no new sales markets were developed. In the early 1970s, the Netherlands' share of cut flower exports declined by five percent as flower producers and exporters from Colombia and Israel entered the market (van Lier 2005, 24). By 1982, long-stemmed roses from the surrounding area and other flowers from further afield were being traded at the Aalsmeer auction (*Nunspeets Nieuws- En Advertentieblad* 1982). Growing conditions are consistently good near the equator due to constant temperatures and hours of sunshine throughout the year.

Analysing at the actors involved along the chain of non-European production, it becomes clear that Dutch influence is central in the major growing regions. This can be illustrated by the cut flower industry in the Naivasha region (Kenya): 'By 1982, nearly 5,000 Kenyans were employed in the Dutch farms' (Porter, Ramirez-Vallejo, and van Eenennaam 2013, 14). Approximately 60% of Kenyan

Fig. 6.20 Visualisation of the spatial structure in period IV and V. (Own illustration)

farms are still part of a Dutch firm or network of individuals. Data collected in Naivasha in 2022 shows that there are often no organisational links between Kenyan and Dutch firms. A significant amount of flower production is managed by or partly owned by Dutch individuals or companies. Some products from Dutch-run farms are imported and distributed exclusively by a firm owned by a relative in the Netherlands. Production in the Netherlands was often discontinued in order to establish production outside Europe. Evidence of this can be seen in the decline in land use for rose production in the Netherlands between 2000 and 2015, when it fell from 932 hectares to 280 hectares (Gemählich 2022, 53). At the same time, the Kenyan cut flower industry around Lake Naivasha expanded from 420 hectares in 1982 (Gemählich 2022, 87–88) to 4,450 hectares in 1990 (Mekonnen, Hoekstra, and Becht 2012, 3727).

In addition, there is a spatial link between non-European regions and various regions of the core through the procurement of production inputs. Breeders, for example, usually have research facilities and showrooms in peripheral production regions (Fig. 6.20).

However, education, research and development activities are still taking place in the Netherlands.

6.4.4 Inherent Power Structure

During period IV, the inherent power structure of the cut flower commodity chain underwent significant changes driven by the increasing globalisation of production and trade, with political power expanding beyond the Netherlands.

The relocation of production processes to regions outside Europe, such as Kenya, meant that political actors in these regions became more influential. Furthermore, the collaboration between the Kenyan government and the Dutch Ministry of Development Cooperation in the early 1980s to promote the Kenyan flower industry exemplifies the expanded sphere of influence of Dutch actors (Porter, Ramirez-Vallejo, and van Eenennaam 2013, 14). The active support of Dutch actors in e.g. Kenya by other Dutch actors, shifting the focus from development aid to economic partnerships, highlights the ongoing yet evolving political influence.

The regulation of market access (political power) also became more internationalised. In 1994, flower growers based in the Netherlands attempted to impose import restrictions on foreign flowers because they feared the negative impact of their increasing presence on prices. In response, the East African Flower Group established the Tele Flower Auction in 1995, which became a successful online platform for international growers. In response, FloraHolland introduced an online version called 'Kopen op Afstand' and eventually took over the Tele Flower Auction in 2010 (van Heck 2021, 97). In 2006, the Dutch auction cooperative opened its doors to international members, including non-Dutch producers from outside the EU (Gebhardt 2014, 100; van Heck 2021, 92). This allowed foreign producers to become full members with voting rights. Full membership allowed them to sell their goods at the flower auction on better terms and to participate in its decision-making process. This has benefited Dutch producers in countries such as Kenya and Uganda (Steen 2010, 116).

These processes demonstrate how the flower auction in Aalsmeer adapted to the evolving global production landscape, integrating international producers in order to maintain its dominant position. Nevertheless, non-European production also strengthened sales and thus the auction's position. In 2008, more imported than Dutch products were traded at the Aalsmeer flower auction for the first time (Gebhardt 2014, 217).

By relocating production to peripheral regions, cut flower producers were able to reduce their production costs, including labour and energy costs. This gave them greater control over pricing and, in theory, enabled them to increase their profit margins. Both interview data and Gemählich (2022, 121) show that 'in the early years of the Kenyan flower industry, profit margins were enormous (up to 50%)'. However, Kenyan roses generally fetch lower prices at auction than Dutch roses. This is partly due to the type of cut flowers produced there.

6.5 Period V: Standardisation (1990s Until Today)

Period V is characterised by a cut flower market distinguished by certified production, standardised products, and market diversification. This process runs partly in parallel with period IV and is the result of European food retailers entering the commodity chain.

6.5.1 Historical Context

Period IV and period V overlap from the 1990s to the 00s. Therefore, I will only briefly discuss the historical context from the 2010s onwards.

In period V, the Netherlands was influenced by various global socio-political shocks, both in terms of direct economic relations and as a central hub for goods.

At the same time, however, period V is characterised by relative political stability and economic growth within the Netherlands.

The referendum on the UK's withdrawal from the EU, held in June 2016, marked the beginning of Brexit. Consequently, the UK has not been part of the EU's single market or customs union since 2020 (Gasiorek and Tamberi 2023, 708). The announcement of the withdrawal and its implementation both had an impact on the European and international economies. This was due, among other things, to the reintroduction of border controls between the EU's external borders and the UK (Gasiorek and Tamberi 2023, 708). The UK remains a key trading partner for the Netherlands. In 2016, the Netherlands exported 23% of its cut flowers to the UK. However, by 2020, this figure had fallen to 18.3%, and by 2022, it had dropped further to 12.5% (The Observatory of Economic Complexity. n.d.-c.). Also in 2020, the whole world was shaken by the rapid spread of the COVID-19 virus. In response to the pandemic, numerous measures were introduced to contain the virus and support local economies as the situation progressed. I have discussed these elsewhere, comparing the export-oriented agricultural sectors in Naivasha, Kenya, and the O'Higgins Region of Chile (Espeter and Retamal 2024). These measures impacted the Dutch economy, in part due to reduced air traffic.

The Netherlands' GDP grew from USD 2,161,000 in 1990 to USD 5,353,000 in 2020. This means that GDP per capita increased almost two and a halffold within thirty years ('IMF 2025). The Dutch economy was in recession in both 2009 and 2020 due to the financial crisis, the pandemic and Brexit (O'Neill 2025a). The Dutch economy's goods trade balance amounted to 58.34 billion USD in 2015, rising to 79.48 billion USD in 2020. It then fell from 82.65 billion USD in 2021 to 67.28 billion USD in 2022 ('Netherlands—Trade Balance of Goods from 2013 to 2023' 2025). Therefore, the impact of the COVID-19 pandemic on the Dutch economy as a whole in the years 2020–2022 can be considered negligible.

The GDP in the North Holland region increased steadily between 1995 and 2022, rising from €66,160 million to €204,169 million, with only a slight dip in 2020 (Statista Research Department, n.d.). In 2022, North Holland remains the province with the highest GDP per capita in the Netherlands, at €70,424 (Statista Research Department 2025).

In 2015 and 2020, the Dutch agricultural sector accounted for a very small share of the national economy (1.81% and 1.74%, respectively). The service sector generated the majority of GDP (just under 70% in 2020) ('Netherlands—Distribution of GDP across Economic Sectors2013 to 2023' 2025). The same applies to the distribution of employees by sector. In period V, approximately

2% of employees were in the agricultural sector (O'Neill 2025c). The number of people employed in the province of North Holland has declined from 39,187 in 2000 to 27,906 in 2020 (Centraal Bureau voor de Statistiek (CBS) 2021). Meanwhile, Aalsmeer's population grew from just under 30,000 in 2010 to just under 32,000 (Netherlands Atlas. n.d.).

6.5.2 Structure of Actors

The structure of actors in period V is characterised, on the one hand, by cut flower producers located in core and peripheral regions. On the other hand, there has been a diversification of distribution channels, resulting in the emergence of new actors. Cut flowers continue to be traded via Royal FloraHolland, which distributes them to end consumers via wholesalers and florists. However, food retailers are increasingly selling cut flowers to end consumers. They purchase the cut flowers directly from the producers, either via Royal FloraHollands's direct trade platform or via import/export companies acting as intermediaries. Royal FloraHolland handles sales and handling in Aalsmeer, as well as financial processing, for members of the auction cooperative. I have carried out a more detailed reconstruction of this commodity chain during for the year 2022 in another publication (i.P.) (Fig. 6.21).

6.5.3 Spatial Structure

The spatial structure of period V differs from period IV in that Aalsmeer is becoming an increasingly important transhipment point in the cut flower commodity chain, rather than a marketplace for them. Direct trade is becoming increasingly important in the cut flower commodity chain. This means that, in period V, only around 50% of cut flowers are auctioned. Nevertheless, the Aalsmeer flower auction remains central, with cut flowers from all over the world being unpacked, inspected, commissioned and shipped from there. However, it is not only the flowers themselves that continue to pass through Royal FloraHolland. Royal FloraHolland also handles financial transactions.

However, direct trade in cut flowers is most likely focused on the food retail sector. Roses with small heads, such as those produced around Lake Naivasha, are primarily intended for this actor. 'The specialised European value chain for high-quality cut flowers still relies heavily on South American suppliers, especially Ecuadorian roses' (Gemählich 2022, 53). The production of roses with larger

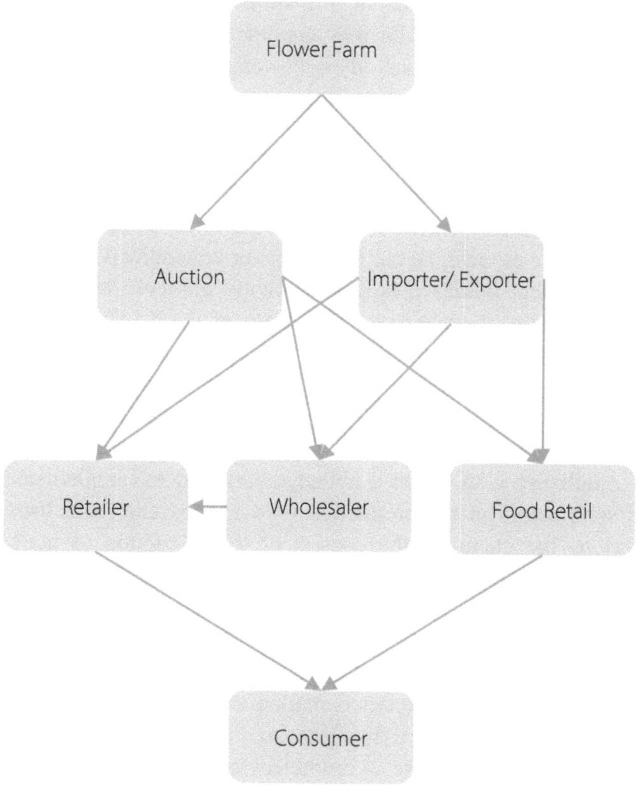

Fig. 6.21 Visualisation of the reduced structure of actors in Period V. (Own illustration)

heads, such as those found in the higher price segment, requires higher altitudes. These are mainly produced in South America, but also in the Netherlands (e.g. Red Naomi), and are sold via the auction system.

6.5.4 Inherent Power Structure

The regulation of processes (political power) is identified as a central element in period V. These are carried out by state actors in the form of occupational safety and environmental protection laws. Furthermore, supranational actors such as the

EU provide the framework for labour migration in the EU area through the EU Freedom of Movement Act. Collective agreements concluded within the commodity chain, such as those observed on flower farms or in the Dutch logistics sector, are also considered a form of process regulation.

These processes can be identified in economic policy measures, such as control mechanisms (taxes, subsidies), as well as legislation (labour and environmental protection), but also in passivity (absence of regulations). The utilisation of financial mechanisms as a means of political power is evident in the provision of economic incentives, such as tax reductions or streamlined lending processes. This form of political influence is predominantly observed among state actors. Beyond the utilisation of governance as a form of political power by state and supranational actors, economic actors also employ this mechanism.

A closer examination of the processes described reveals that the instruments of power differ in their influence. Instruments of political power have an effect on different processes along the chain. For example, European plant protection measures, compliance with which is already controlled and implemented by economic and state actors in the production of seedlings and from then on again and again along the chain. In the context of the restriction of market access, this aspect constitutes a component of the political power exercised by state and supranational actors.

Nevertheless, the restriction of market access is a fundamental part of the inherent power structure of the cut flower commodity chain. On the one hand, the flower auction in Aalsmeer restricts market access. The marketplace operated by the cooperative is only partially open to the public, with access requirements decided by the cooperative's committees and the members they represent. Although non-members can sell their goods at the auction, they have to pay a percentage of the sale to the auction. Furthermore, the auction establishes quality standards for the goods, which are binding and subject to random inspection. For instance, the Royal FloraHolland has determined the degree to which the buds of flowers must be opened, serving as an indicator of vase-opening potential, even when separated from the mother plant. Additionally, all sellers are obligated to provide information regarding their products, a requirement that extends to both direct sales and auctioning. The requisite information encompasses colour, variety, head size, stalk length, quality level (indicating the degree of openness of the flower), bundle size, and, given the predominance of online sales, photographic documentation of the goods. This service is available for purchase through the auction.

Political power is also derived from the certificates, which play a significant role in period V due to their impact on potential market access. All the farms under study have at least one, and frequently several, of the certificates commonly utilised in the chain, including certifications from MPS (certification organisation established by the Dutch flower industry), the Kenyan Flower Council (KFC—lobby organisation of the Kenyan flower industry) and FairTrade (public-benefit organisation). These certificates are issued by entities operating within the neighbouring systems of the commodity chain. The significance of these certificates is particularly pronounced in the context of direct trade, either via or bypassing the flower auction.

FairTrade-certified roses, for instance, have become a prominent component of the product range offered by European for retailers, underscoring the importance of these certifications for the farms seeking access to specific target markets. The process of obtaining these certifications involves an audit, which necessitates financial investment and must be periodically renewed. The interviewees describe this process as time-consuming and costly, but also as necessary for participation in the market. Furthermore, the time required is manageable once an actor has already gone through and received certification, since the requirements are very similar and include occupational safety, environmental protection, and quality standards. The scope of certification varies depending on the specific certificate; some certifications apply to all the flowers harvested on the farm (MPS & KFC), while others apply only to a predefined quantity (FairTrade). Additionally, the criteria for certification differ depending on the certificate type. In some cases, certification is required for the farm only, while in others, it is required for all parties involved in the commodity chain of a particular variety, such as logistic, import/export, or retailers. The use of the corresponding label is only permitted for those parties who have obtained certification (Fig. 6.22).

Within the context of the commodity chain, economic power is manifested through the control exerted over resources and means of production, thereby enabling the generation of added value. Initially, the cost-intensive rose production in Kenya was mainly financed by European parent companies, as it was difficult to obtain loans (Owner 2022a).

The European food retail is characterised by relationships between purchasing associations, with the four largest purchasing associations in terms of revenue being European Marketing Distribution (including Kaufland), Coopernic (including REWE Group, Coop Italia), Epic Partners (including EDEKA, Jumbo, Picnic) and Agecore (including Coop Switzerland, Kaufland Germany) (Largest food purchasing groups in Europe 2021, 2024). It is notable that all four groups have their headquarters in Switzerland. In addition to these purchasing associations, there are florist shops at on end of the chain, though these are mostly craft shops

Fig. 6.22 Certified and uncertified cut flowers in German supermarket (left) and discounter (right). (Own picture, 2024)

with few employees. In recent decades, florist chains such as Blume2000 in Germany and the growing online trade have been added, though the latter plays a subordinate role in the purchasing behaviour of consumers (Schirmer 2024).

The cut flowers are sold approximately 50/50 in direct sales and through auctioning. However, seasonal sales at Valentine's Day, Mother's Day and Women's Day have a significant impact on pricing. These sales peaks are used in the negotiations of contracts for direct sales, but do not represent power in the commodity chain. Rather, it is the reputation of the cut flower producer that exerts significant influence on its economic power. This reputation has a substantial impact on pricing across both auctioning and direct sales channels. The purchasers, in turn, rely on the reputation and their past experiences with the flowers from the farm or the products from related companies. Unlike, for example, at the fish market examined by Dobeson (2016). The reputation of the producer is reflected in the prices set by the auctioneers at the beginning of the auction (Auctioneer 2022).

In contrast, direct trade is characterised by a distinct market environment, one that is markedly different from sales conducted via auction houses. This disparity in market power is reflected in the bargaining position of buyers, who, due to their market dominance and the influence of purchasing associations possess a significant negotiating leverage in terms of favourable contractual conditions. For instance, orders are placed well in advance, often several months prior to the actual delivery. Market prices at the flower auction in Aalsmeer serve as a benchmark for price negotiations, providing a reference point for the negotiation process. The role of import/export companies as intermediaries between buyers and sellers is crucial, and these relationships are often dependent on maintaining positive connections. These companies operate in both direct trade and through auctions.

Furthermore, a substantial network of personal relationships has been identified within the cut flower commodity chain (Gebhardt 2014). Longstanding contacts have been observed between buyers and cut flower producers, with the origin of the farm management playing a significant role. This is often reflected in the products' target markets. Companies of European origin often focus on the European market, whereas others also export some of their products directly to Asian countries or the US market.

The bargaining power of producers is reflected in their share of the selling price. For most of them, these are rising fixed costs. At the same time, market prices are barely increasing. According to Gemählich's exemplary cost breakdown (2022)—derived from the balance sheets of various flower farms—the share of costs for a rose (retail price 37 euro cents) is as follows: retail in Europe 24.1 cents (65%), production in Kenya 4.9 cents (14%), including production costs (2.5 cents), labour (1.8 cents) and profit (0.6 cents). Distribution and auctioning account for 8 cents (22%). In the Netherlands, therefore, 8 cents are earned per stem, while just under 5 cents are received during production. On the basis of the underlying data, no statement can be made about the cost of certificates. However, the need for them to participate in the market is impressively demonstrated. As a result, the issuing institutions have a strong position of in the inherent power structure of the cut flower commodity chain in period V. The positive impact of certificates such as FairTrade on workers and their environment can be seen in older reports from the Naivasha region (Hale and Opondo 2005; Omosa, Kimani, and Njiru 2006) and in current reports on production in Ethiopia (NDR 2023). However, a more comprehensive examination of these data sources in combination with the interview data already collected is needed.

With regard to cut flower production, it should be noted that non-European producers, such as those in the Naivasha region, largely supply the food retail trade. However, standardisation is important for these producers in terms of more than just market access. The transportability of cut flowers, which is determined by factors such as shelf life, weight and size, also influences market prices.

7 Analysing Commodity Chains in the Longue Durée

The aim of my book was to make the process-nature of economic structures analytically tangible by focusing on how power is established and maintained within them. I introduced the process-oriented commodity chain mapping as methodological approach to collect data on and analyse economic structures by applying world-systems analysis (Wallerstein 2011a [1976]; 2011b [1980]; 1989; 2011c) and the concept of commodity chains (Hopkins and Wallerstein 1977; 1994a).

I used the modern world-system as a unit of analysis. This illustrates that the subject of my study is a stucture that exists through economic relationships. These relationships structure the world-economy through market exchange, the division of labour, and the relationship between states, firms and their shared goal of maximising capital accumulation. Applying the concept of commodity chains by Hopkins & Wallerstein to the modern world-system is useful for understanding and analysing its economic structure. Commodity chains describe the structure of the world-economy. They focus on the connection between production processes, thereby reflecting the axial division of labour within the system. Hopkins and Wallerstein's theoretical concept is appealing in its simplicity and was therefore extremely well suited to this book.

My empirical study involved reconstructing the cut flower commodity chain and focusing on three key structures: structure of actors, spatial structure, and inherent power structure. Each structure was analysed within the previously identified periods, providing snapshots of the cut flower commodity chain. Adopting a longue durée perspective enabled me to trace the process-nature of this chain from the formative phase to 2022, identifying historical and recent processes. While analysing the data I distinguished between relevant and less relevant actors and addressed the challenges of historical reconstruction, such as limited data availability. To ensure comparability along the longue durée, certain components, such as input materials, were omitted.

L. M. Espeter, *From Hobby to Billion-Dollar Industry*, Wirtschaft + Gesellschaft,
https://doi.org/10.1007/978-3-658-52437-1_7

In addition to their theoretical concept of the commodity chain, Hopkins and Wallerstein set out a methodological approach to researching commodity in the longue durée. Inspired by their work, I have introduced process-oriented commodity chain mapping for examining and analysing the process-nature of economic structure. Through this and the subsequent empirical study, I demonstrated how commodity chain research in the longue durée can be applied.

Unlike Hopkins, Wallerstein, and other scholars, I divided the research period into meaningful periods. In order to overcome limitations, this approach allowes for flexible, question-oriented periodisation, which I refined throughout the research process. Thereby, I placed the process-nature of the structure at the analytical centre right from the beginning. Through conducting my case study of the cut flower commodity chain and integrating the work of Hergesell, Baur and others, I determined the formative phase 'Imperialism and Curiosity' using backward reading. Taking the underlying research question into account, I then applied forward reading to identify the content-driven periods: 'Period I: Start of Commercial Cut Flower Production (1850s–1910s)'; 'Period II: Establishment of Central Locations of Trade (1900s to 1950s)'; 'Period III: Economic Growth and Centralisation (1950s–2000s)'; 'Period IV: Repositioning Within the Chain (1980s–2000s)'; 'Period V: Standardisation (1990s until today)'. Although the processes forming the periods run partly in parallel, the periods still form a chronology. Each period has a before, a during and an after.

In the second analytical step, I reconstructed the structures of the cut flower commodity chain. Considering the product properties, I reconstructed the structures within the meaningful periods identified. In the reconstruction of the structure of actors, I identified the relevant actors and their position in the commodity chain. This provides a general insight into both the period and the structure. Rather than focusing on individual production steps, this simplified representation focuses on the relationships between the actors. It therefore provides a basis for analysing the spatial and inherent power structure. However, during the analysis of the data, it is necessary to distinguish continuously between actors that are relevant to the research question and those that are less relevant. In addition, the availability of data poses particular challenges in historical reconstruction. For this reason, various components of the commodity chain have been omitted from the reconstruction to ensure comparability. This applies in particular to input materials, such as greenhouse technology or fertilisers and pesticides. Based on this, I then reconstructed the spatial structure, and finally the inherent power structure. The latter was based on an operationalised definition of power based on Marx, Weber, and Wallerstein. These reconstructions enabled me to depict the cut flower commodity chain as a series of moments in time.

In the third analytical step, I examined the process-nature of the cut flower commodity chain within and across the meaningful periods, based on the reconstructed structures. This involved analysing the cut flower commodity chain in the longue durée. I analysed it from its formative phase through to five subsequent periods, up to 2022. By focusing on the cut flower sector in Aalsmeer, I examined the internal and external processes of the commodity chain. This is despite major upheavals and the spatial restructuring of the commodity chain being identifiable since the post-war period. Applying a longue durée approach, I examined long-term processes, identifying connections and correlations between them and their role in establishing and maintaining power within the economic structure of the cut flower commodity chain.

7.1 Expanding World-Systems Analytical View

Although I identified processes in the empirical data similar to the Kondratiev cycle, which form the basis of periodisation according to Hopkins and Wallerstein, periods of contraction and expansion did not cause major upheavals in the cut flower commodity chain. For instance, the global economic crisis of the 1930s impacted the Netherlands, including the Aalsmeer region and its cut flower industry in period II. Sales at cut flower auctions in Aalsmeer halved. At that time, the state provided financial support to the sector in the municipalities of Haarlemmermeer and Aalsmeer. However, this did not significantly alter the structure. Instead, the state supported the already dominant position of actors within the commodity chain, and thus of the flower auctions in Aalsmeer. The time of the Second World War and the German occupation of the Netherlands can be described as a pause for the cut flower commodity chain. The war led to a collapse in general economic activity and trade. However, the previous structures of this commodity chain were revived in the post-war period. This period brought general economic growth in period III. This growth significantly impacted the cut flower commodity chain and led to the merger of the flower auctions in Aalsmeer. Nevertheless, it can be argued that economic growth was not the sole factor in altering the structure. First, the CAV lacked the space to expand accordingly. In addition, the position of the CAV and Bloemenlust within the power structure was undermined by new cut flower producers and auctions. Therefore, the merger to form the VBA represents a process of maintaining and expanding the power of the Aalsmeer flower auctions.

The energy crisis of the 1970s in period IV was a phase of contraction, with probably had the greatest impact on the cut flower commodity chain. However, this was not due to a lack of sales. In fact, the Aalsmeer flower auction expanded its area significantly during this period, continuing to do so until the 1990s. Instead, following the energy crisis, oil, a vital component in the production process, became significantly more expensive. During period IV, it became apparent that production processes were relocated to peripheral regions where this input material was less necessary (i.e. regions where greenhouses were not artificially heated or lit). These peripheral regions of the modern world-system also have different production conditions and costs compared to those defined as core. The spatial and functional reorganisation of actors within the cut flower commodity chain can be attributed to the contraction phase following the energy crisis. Nevertheless, the relocation of the production processes of cut flowers, which are a very fragile commodity, probably only became technologically and politically feasible in period IV.

Based on the data I collected and analysed, I can hardly detect any impact of the 2008 financial crisis on the cut flower commodity chain. The only notable circumstance is that sales at the flower auction in Aalsmeer remained stable during period V and only increased again afterwards.

In summary, I have been able to show with this book that a large number of internal and external processes have influenced the structure of the cut flowers commodity chain. But the general phases of contraction and expansion in the European economy had little influence. Rather, processes of power establishment and maintenance shaped the structure. Therefore, dividing the study period into meaningful periods has proven useful for examining the process-nature of economic structures in the longue durée.

7.2 Relationship Between State and Firms in the Cut Flower Commodity Chain

In my analysis of these processes, I examined that the Dutch state does not merely promote individual firms. Rather, the case of the cut flower commodity chain clearly demonstrates that state actors promote entire sectors, branches of the economy, and even regions. For instance in period II and III, during the financial crisis of the 1930s and the merger of VBA and FloraHolland, the Dutch state prioritised the flower auction in Aalsmeer and the surrounding arctors. This region and the flower auctions in Aalsmeer already held a monopoly position prior to state intervention, a position that was then secured and promoted by the state. Furthermore, the Dutch government has been promoting the cut flower commodity chain through state-funded education and research institutions since period I.

In addition, Aalsmeer benefited from major infrastructure projects such as the Mainport Strategy and the expansion of Schiphol Airport into one of Europe's largest passenger and commercial airports, which made it a prime production and trading location. However, during Periods I and II, the cut flower industry based in Aalsmeer already benefited from government-funded infrastructure projects, such as the expansion of the canal system, road and rail networks. Since period III, the preference for the flower auction in Aalsmeer becomes evident in traffic planning being oriented towards auction times. In Wallerstein's concept of the capitalist world-economy, the relationship between states and firms are of central importance. Sovereign states can implement protectionist policies to support firms in line with their position within the interstate system. Through these measures' states can promote firms' economic activities and the associated accumulation of capital

7.3 Axial Division of Labour in the Cut Flower Commodity Chain

The concept of axial division of labour, in the original sense of a spatial and functional division of production processes, only came to fruition during period IV. Previously, the cut flower industry in the Aalsmeer region was organised around production, trade and logistics. However, with the shift in production, the commodity chain has become more functional and spatial. While cut flowers are increasingly cultivated in peripheral regions, research and development, trade, distribution and logistics are centred in core regions around Aalsmeer. In this process, the production of cut flowers has become a 'peripheral box' period IV is therefore pivotal to the axial division of labour and the resulting unequal exchange.

However, the spatial expansion of the cut flower commodity chain can be observed even before then. In terms of the division of labour along the chain, this means that rising production costs in the core have led to the shift of production processes, thereby integrating peripheral regions into the economic structure and providing new labour and sales markets for other core-like products. I demonstrated that the integration of peripheral regions does not only occur through the relocation of production processes. Instead, the cut flower commodity chain was opened up to cheaper labour as early as period III through illegal and semi-legal labour migration. Furthermore, various actors along the commodity chain have benefited from the free movement of workers within the EC and EU since the late 1950s. The geographical expansion of the EC/EU has simplified relations between regions with similar production processes and the availability of cheaper labour, for which different labour rights apply. This specific form of labour integration in the German agriculture is examined empirically for example by Griesi with regard to the organisation of labour recruitment (i.P.) and Boatcă regarding exploitation during the early COVID-19 pandemic (2020).

A similar pattern can be seen after the relocation of production processes. In the Lake Naivasha for instance, there has been domestic labour migration. The majority of workers on cut flower farms come from the Bungoma, Kakamega and Kisii regions (Kuiper and Greiner 2021, 7), rural regions in the west of Kenya. According to their own statements, they plan to return there after their employment. These workers are cited as an advantage for Naivasha's location over that of other regions, such as Ethiopia (Upper Management 2022a). This is because the 'value creation is more and more based on a stable, experienced, and trained workforce, which has potentially positive effects on labour conditions' (Gemählich 2022, 146).

My analysis shows that a shift in production processes is not a prerequisite for the axial division of labour. Instead, processes such as interstate agreements and internal migration ensure that cheap labour-power is available for production.

7.4 Spatial Expansion of the Cut Flower Commodity Chain

The original aim of the spatial reconstruction was to focus specifically on the axial division of labour, as outlined by Hopkins and Wallerstein. However, my study revealed that this reconstruction also shed light on other aspects of the spatial expansion of a commodity chain, such as land acquisition, spatial limitations in urban areas and the expansion of the sphere of influence through alliances of states and supranational organisations.

Based on my research findings, I argue for closer examination of the significance of spatial expansion in the world-economy beyond the axial division of labour. My analysis of the cut flower commodity chain revealed numerous processes of spatial expansion, e.g. the access to natural resources and productive lang.

However, this does not refer to the original sense of the term, as no production processes were relocated or linked together. Even in the formative phase, it became apparent that the spatial expansion of the European and Dutch sphere of influence through imperialism had a considerable impact on various processes in Dutch society. For instance, it impacted the self-image and identity of Dutch

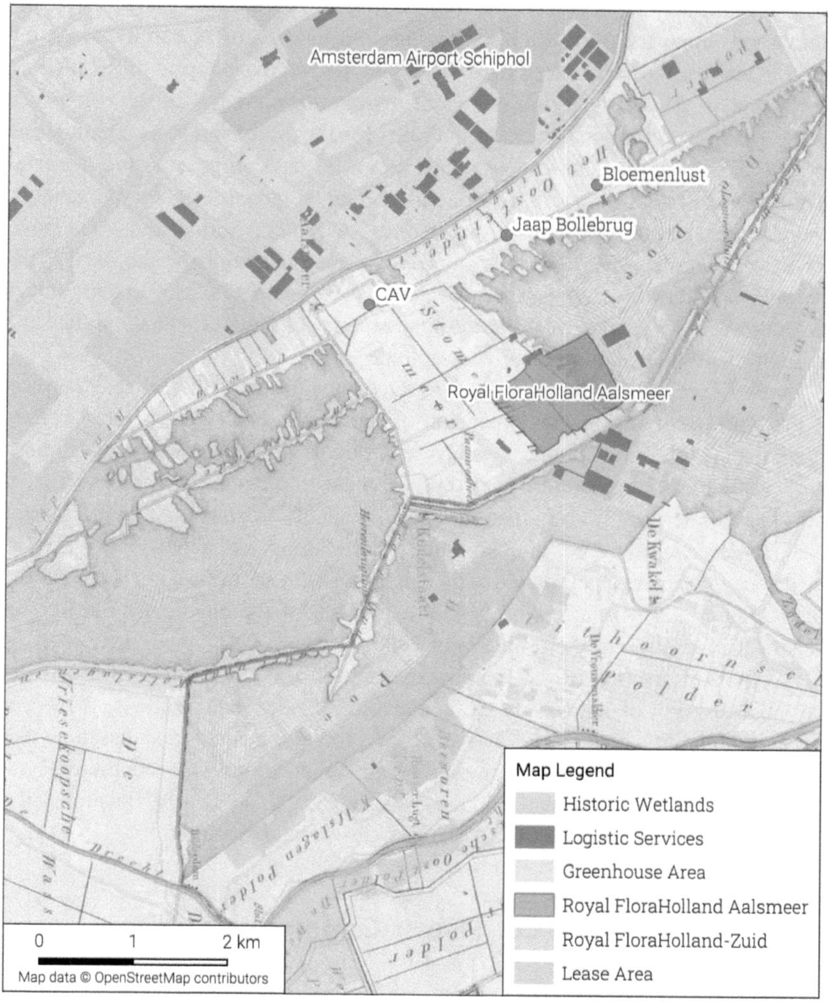

Fig. 7.1 Cut flower industry around Aalsmeer illustrated map of 1860. (Own illustration, created by Moll)

society as a trading society. Or the interest in new and unusual things that arose from goods imported from colonised regions. Ultimately, my research into the cut flower sector can also be traced back to the initial spatial expansion of the modern world-system. This is therefore of central importance not only for the emergence of the cut flower commodity chain, but also for the Netherlands as a nation and society. This phenomenon can probably be traced back to 'socialising agencies of the world-system', such as households and state agencies, including schools. These shape and convey 'status-group identity' within a world-system. 'The powers [...] hope that socialization results in the internalization of the myths, the rhetoric, and the theorizing of the system' (Wallerstein 2006, 37). The VOC's violent undertakings also established the prosperity of Dutch traders and the state through their spice and enslaved person trade.

Furthermore, in this book I demonstrated that the further spatial expansion of the world-economy in the expansion of productive land within state borders. In particular, draining moors, swamps and larger bodies of water made these areas available for agriculture and thus capitalisation. Initially, the drainage of the Haarlemmermeer meant that the population of Aalsmeer could no longer fish. However, agriculture, and eventually horticulture, could later be practised on the newly created peaty soil, which is the origin of the region's cut flower sector. Figure 7.1 visualises the spatial expansion of the cut flower industry in Aalsmeer in relation to the usable land 1860.

Another observation of the spatial expansion of the world-economy—beyond the axial division of labour—is the connection of new sales markets to the cut flower commodity chain. This has enabled production and trade based in the Netherlands, as well as the flower auctions in Aalsmeer, to grow continuously.

Through its examination of longue durée, this book clearly illustrates the various ways in which economic structures can expand spatially.

7.5 Power in the Cut Flower Commodity Chain

To analyse the power structure of the cut flower commodity chain, an explicit definition of the concept was necessary. For this purpose, I incorporated Marx's definition of power, which determines access to economic and political resources. The definition is convincing in its simplicity. Furthermore, world-systems analysis is largely influenced by Marx's work. I operationalised political and economic power with regard to the commodity chain as the research subject. This definition and operationalisation of power proved crucial in identifying processes within the cut flower commodity chain.

However, empirical evidence showed that collective power within the structure is also significant, in addition to the various types of political and economic power, for example, through the ability to influence pricing or market access. This enables actors to occupy their positions differently (e.g. through cooperation or official alliances). In future work, Hannah Arendt's concept of power could be used to supplement this from the beginning. Arendt understands power as something that emerges from a collective. Collective action is based on consensus and cooperation (Arendt 1970, 44).

Using the previously defined and operationalised concept of power to reconstruct the inherent power structure has proven particularly useful for examining the processes-nature of commodity chains. As mentioned above, I identified the establishment and maintenance of power as central to the processes-nature of economic structures. The actors under consideration reacted to these processes by primarily taking collective measures to secure or expand their power.

7.5.1 Individual and Collective Interests of Power

The process-nature of the cut flower commodity chain can be traced back to endeavours to establish and maintain power. These efforts can be divided into three categories: individual interests of power (i.e. individual actors); collective interests of power (e.g. associations and cooperatives); and of both. Thus, the promotion of monopolisation can be attributed to state actors.

Many of the efforts observed within the commodity chain can be summarised as attempts by companies to pursue individual interests of power. Examples include relocating individual production processes from the European core to the non-European peripheral. Another example is the shift in a firms' focus from production to research and development. Furthermore, the installation of refrigeration facilities at Kenyatta Airport in Nairobi, Kenya, is another example. This measure was initially taken by a flower farm's individual interest, but subsequently benefited other farms. Another example is cooperative members bypassing the auction as a place of sale. However, the collectivisation of the interests of power is crucial. Actors within the cut flower commodity chain have established and maintained their political and economic power by forming company mergers and alliances, and by joining auction cooperatives. These processes all aim to establish or maintain individual or collective positions in the power structure.

The process-nature of the cut flower commodity chain can also be traced back to a common phrase: 'God created the world; the Dutch created the Netherlands'.

Even in the examined formative phase, collective Dutch ventures can be identified that have shaped the actors, spatial, and inherent power structure of the cut flower industry in the longue durée. On the one hand, there is the network of flower enthusiasts. The research and trade in flower bulbs carried out by this network formed the basis of the commodity chain. The importance of this network's contracts and involvement in overseas trade and imperialism is evident. They also played a role in technological projects such as the 'Beemster Polder'. Personal networks and trust characterise the actor structure of the cut flower commodity chain. These *collective* ventures made commercial cut flower production possible in the first place.

However, the transfer of firms and knowledge across generations within families can also be interpreted as a process of consolidating collective power.

When focusing on the actor flower auction, it is challenging to distinguish the collective interests of the cooperative members from the individual of the auction cooperative: The establishment of the flower auction cooperatives in period II from the collective interest of flower growers in improving their political and economic power. The merger of these cooperatives during period III strengthened the position of the farmers and the cooperative themselves. In period IV, integrating farms in peripheral regions into the cooperative enabled this actor to extend their spatial influence. It also enabled the flower auction(s) in Aalsmeer to maintain their position within the commodity chain. The auction cooperative and its members benefit from the handling of financial transactions, including direct trade, by the auction. The latter benefit from receiving reliable and quick payment for their goods. The former benefit from a constant flow of money through their organisation, including percentage shares. According to Gebhardt, the flower auction 'FloraHolland seeks to protect and defend its interests, which don't always necessarily coincide with the interests of Dutch growers' (Gebhardt 2014, 110).

Furthermore, the organisation of the cooperative itself is constantly changing. The establishment of offices and departments, including marketing and digitisation (van Lier 2005), is intended to improve the auction's own position or that of its members. This constant internal process serves the interest—of the cooperative and its members—of maintaining and improving the position of power within the cut flower commodity chain.

7.5.2 State Protectionism in the Cut Flower Commodity Chain

A particular form of interest in power and the ability to enforce it is that of the state. Some of these actors are democratically elected representatives of the people, while others are administrative employees.

The Dutch cut flower industry is closely linked to state actors and their protectionist measures. As early as the mid-19th century, state projects such as 'reclamation projects', which aimed to reclaim land for agriculture, paved the way for the expansion of agricultural land (Wintle 2000, 113). At the same time, the enclosure of common land between 1820 and 1860 contributed to the intensification of agricultural use. This process continued with the Enclosures Act (Markenwet) of 1886, which was followed by further land development in the east, largely driven by the Netherlands Heath Company (Nederlandsche Heide Maatschappij), which was founded in 1888. The company initiated a campaign to convert heathland and other common land into productive farmland. The continuous support of the state has been a decisive factor in the growth and success of the Dutch cut flower industry. As previously demonstrated, the industry's exceptional position in the Dutch economy is inextricably linked to the country's history and identity, spanning everything from imperialism and water and land management to scientific developments. The cut flower industry has become an integral part of modern-day Dutch society. Key actors such as the flower auction cooperative in Aalsmeer are deeply embedded in state policies, national discourse, and historical circumstances. The Dutch business elite and the upper echelons of horticultural networks are closely intertwined playing a decisive role in the political landscape.

Another significant contribution made by the state is the promotion of research and development. Although the exact dates of the establishment of individual institutes vary, horticultural research was established during the formative phase and received state funding during period I. Today, numerous research institutes, projects, groups and study programmes are dedicated to plant breeding, cultivation conditions and related areas. Additionally, an established system of vocational training ensures a supply of skilled workers.

Furthermore, from period II onwards, it becomes clear that the Dutch cut flower industry is supported by state actors. The fact that state support is intended to benefit both the Dutch flower industry and the Dutch economy as a whole is illustrated by the example of KLM and the flower industry. Securing affordable, efficient transport routes is important not only for the Dutch flower industry, which acts as a global trade hub, but also for the strategic interests of the Netherlands as an export nation. For East African flower producers, who are heavily dependent on air transport, flight costs can account for up to 40 per cent of sales value (Gebhardt 2014, 214). The competitiveness of Kenyan and Ethiopian producers is significantly influenced by the availability and prices of air transport, with KLM and Dutch government support playing a decisive role. This strategy becomes even clearer when considered in the context of the European internal market. The Netherlands benefits significantly from open markets and the free movement of goods within Europe. By promoting key industries such as the cut flower trade, the government strengthens the economy and consolidates the country's position as a central logistics and trade hub in Europe and globally. Supporting the cut flower industry is therefore a means to an end, helping the Dutch government to achieve its overarching goal of establishing the Netherlands as a leading trading nation.

However, it is important to note that the success of the Dutch cut flower industry is not solely due to government protection. Van Zanden (1987) points out that agricultural performance was particularly strong in sectors that received little or no protection in the 1950s and 1980s.

Nevertheless, the Dutch state's clear influence on the cut flower industry shows that it did more than simply support or promote individual companies. Instead, it promoted firms and their personal networks by supporting the Dutch part of the commodity chain as a whole. In doing so, it strengthened its own position within the modern world-system. The capital interests of those involved in the commodity chain and the state's interests go hand in hand. While individual cases do not involve millions (most people earn a moderate income), a significant number of jobs and tax revenues are at stake in the Netherlands overall.

7.6 Continuity in the Cut Flower Commodity Chain

'Where do cut flowers come from? The Netherlands.' This was the answer, or something similar, that I heard time and time again from a wide variety of people while working on this book. At least among end consumers and partly in the wholesale sector. While this statement is just one example, it nevertheless provides a good description of the perception of cut flowers as a commodity and the commodity chain.

Examining the cut flower commodity chain in the *longue durée* reveals the consistent presence of *Dutch* actors. This is evident both over time and along the commodity chain. The commercial cultivation and distribution of cut flowers in the Netherlands began with a group of flower enthusiasts and botanist interest in exotic plants, resulting in research. Dutch actors are present in each of the examined periods. This is due, first, to the historical origin of the commodity cut flower, and second, to their involvement in all aspects of the chain over time. This is also evident in production in Brazil, Canada and the USA, where the main growing regions have experienced significant migration from the Netherlands.

Starting with the inputs, such as greenhouses and greenhouse technology, as well as research and development, these have mostly come from Dutch firms in the Netherlands. Turning to production, until period V, a large proportion of global cut flower production took place in Dutch firms in the Netherlands. Even with the relocation of production processes to the periphery such as the region around Lake Naivasha, however, the connection to the Netherlands remains. This becomes evident as the production either takes place in Dutch firms or is largely managed by people of Dutch origin. In addition to internationally active logistics firms, there are also logistics firms with close personnel and organisational links to the Netherlands. These firms specialise in transporting cut flowers. Examples include Bloemenexpress and Airflo. The former belonged to the Aalsmeer flower auction and transported cut flowers by truck throughout Europe. The latter organises the transport of Kenyan cut flowers to Europe for direct sale or auction. Founded by a Dutch owned flower farmer on Lake Naivasha, now part of the Dutch Flower Group.

Since the inception of air transport for cut flowers, Schiphol Airport and KLM have dominated the cut flower commodity chain. Throughout this time, Schiphol Airport has been a central hub in the commodity chain, particularly following the increase in international production in period IV. This has been reinforced throughout all periods by political actors.

Dutch firms and management personnel also dominate the import and export sectors, as well as the wholesale sector. Most firms specialising in the trade of cut flowers or other plants are Dutch. Even at the wholesale flower market in Berlin, Dutch traders act as intermediaries. These actors are linked not only by their shared origin, but also by the fact that they have often been involved in the cut flower industry for several generations. Within the commodity chain, personal (and sometimes family or formerly religious) networks characterised by a high degree of mutual trust exist. Furthermore, this industry is ‘often, through personal networks, […] never more than one or two degrees of separation from national policies, ranging from foreign affairs to infrastructural planning’ (Gebhardt 2014, 16) (Table 7.1).

Table 7.1 Changes within the reconstructed structures from one period to the next. (Own illustration based on own data)

	Period I → Period II	Period II → Period III	Period III → Period IV	Period IV → Period V
Actor Structure	Central actors, such as cooperative flower auctions (e.g. in Aalsmeer) establish	Increasing number of cut flower producers	Companies restructuring internally (e.g. moving production to periphery or conducting research and development in core)	New actors: food retail & certifiers
	Actor structure becomes differentiated	Decreasing number of cooperative flower auctions	new non-European actors	
Spatial Structure	The market for Dutch cut flowers is expanding geographically	Flow of goods is centralising around a few trading centres	Expansion of the sales market	Expansion of the sales market
	Increasing amount of goods are passing through central Marketplaces	Sales market is expanding massively	Expansion of accessible labour-power	
			Relocation of production processes	
Inherent Power Structure	Farmers gain power by pooling their interests	Remaining cooperative flower auction stabilise and expand their power	Declining influence of Dutch actors on production	Food retail establishes power through market dominance
			Increasing influence of supranational actors on production and trade	Import/Export companies gain power through the growing Importance of direct trade

(continued)

Table 7.1 (continued)

	Period I → Period II	Period II → Period III	Period III → Period IV	Period IV → Period V
	Establishment of position of power through cooperation		Integration of non-European actors into the cooperative flower auction Aalsmeer, stabilisation and expansion of power	Certifications are key to market access
				Cooperative flower auction Aalsmeer secures its position through integrating direct and online sales, as well as digital auctions

Additionally, the cut flower commodity chain is male-dominated, particularly at senior management level. Women are found primarily in harvesting roles in peripheral regions, as well as in sales and marketing. Gebhardt describes the cut flower industry as 'dutch variety of old-boy system' (2014, 19–20).

In addition to the personnel and organisational interdependencies, the case also shows a spatial dependency (Kirchner and Baur 2025) for the Aalsmeer region—comparable to region networks, such as those studied by Saxenian (2000) and others.

The area surrounding Aalsmeer is home to a number of firms involved in production, trade, logistics, and research and development. Figure 7.2 maps the actors in the Aalsmeer region geographically.

The only part of the cut flower commodity chain that has never been dominated by the Dutch is the distribution of flowers to end consumers. This has varied across regions and countries over time. In Berlin, for instance, the sale of cut flowers has been dominated by former Vietnamese guest workers in the German Democratic Republic since the 1990s. Since the start of period V, supermarket chains organised in European purchasing associations have become increasingly dominant in the European part of the commodity chain.

The continuity of the cut flower commodity chain therefore lies with the Dutch players and the flower auction in Aalsmeer. They have been an integral part of the chain since its inception. Since then, they have established and maintained their power, influenced by internal and external processes.

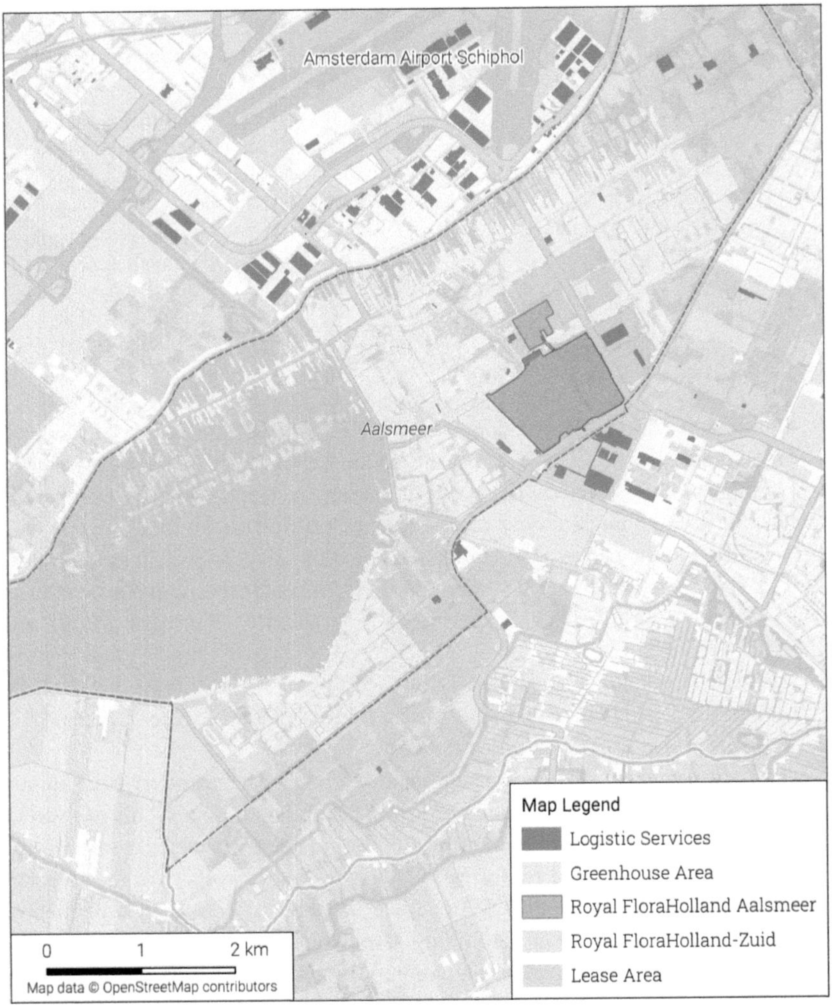

Fig. 7.2 Spatial dependency of the cut flower industry in Aalsmeer. (Own illustration, created by Moll)

7.7 Researching Economic Structures in the Modern World-System

By analysing the cut flower commodity chain, focusing particularly on the cooperative flower auction(s) in Aalsmeer as a central case study, I demonstrated the process-nature of economic structures. A key finding is that space was of central importance. This is evident from the early connection between the VOC and the development of the Dutch cut flower industry. The spatial expansion of the Dutch sphere of influence through the intertwining of trade routes and raw material procurement across national borders demonstrates this early connection. This expansion was largely achieved through the violent subjugation of other societies. Over the period under review, the spatial spread of the commodity chain increased. Key examples include drainage and enclosure measures, access to new markets and cheaper labour through the establishment and expansion of the EC/EU, and the shift of production processes to peripheral regions in the southern hemisphere. These processes should not be viewed in isolation, but rather as integral components of the structure of the capitalist world-economy. Like other processes, these examples are part of the spatial expansion of the cut flower commodity chain. Land, resources and labour are integrated into the chain.

Through these and other empirical findings, I contribute to the theoretical discourse of world-systems analysis and commodity chain research. I demonstrate that the commodity chains approach is not only compatible with, but also applicable to, the modern world-system. My work confirms that spatial expansion, and thus the integration of new regions into the modern world-system, is an integral part of its development. However, I have shown that this does not occur solely through the division of labour and the relocation of production processes. Instead, the world-economy is expanding through the development of commercially usable land (e.g. drainage of lakes and moraines), the introduction of new sales markets (e.g. establishment of the EU's internal market) and the utilisation of new sources of cheap labour (e.g. free movement of workers within the EU). My case study therefore reveals the extremely complex nature of the spatial expansion of the capitalist world-economy. Moreover, my results confirm key assumptions of world-system analysis, particularly the importance of the relationship between states and firms in gaining and securing dominant positions within the modern world-system. Furthermore, it has become evident that, to comprehend the process-nature of economic structures, world-system analysis must be expanded to encompass a precise definition and operationalisation of power. In this book, I have proposed a definition of power based on the work of Marx and Weber. When analysing economic structures, I characterise power as the ability to assert one's own interests within that structure. I distinguished between political and economic power. Political power is the ability to regulate processes and influence political decision-making. Economic power, on the other hand, encompasses access to and control over resources and means of production.

My work contributes to the methodological discourse on investigating economic structures in the modern world-system by introducing the methodological approach of process-oriented commodity chain mapping. This approach combines the methodological concept of commodity chains, as outlined by Hopkins and Wallerstein, with process-oriented methods developed by Baur, Hergesell and others. It builds upon these methods in a targeted manner. Through the introduction of process-oriented commodity chain mapping, I have demonstrated that the systematic and comprehensible collection and analysis of commodity chains in the longue durée is possible and fruitful. This methodological integration provides new insights into empirical research on global commodity chains, offering a more nuanced understanding of complex economic processes over time. The content-based, which is therefore also research question-based, periodisation as analytical step represents a substantial adaptation of the periodisation proposed by Hopkins and Wallerstein. This allows researchers to identify these periods throughout the research process. Consequently, they can constantly evaluate their suitability and make adjustments as necessary. Throughout the analysis, the research question and the case remain the constant focus.

Notwithstanding the findings I presented in this book, my study is not without limitations. The selection of cases is a methodological limitation: the cut flower industry in the Aalsmeer region is both typical and extreme. However, due to its specific historical development, it cannot easily be transferred to other production regions. Nevertheless, processes along the longue durée identified provide important insights into the interactions between the power interests of economic actors, their influence on other actors and spatial structures, and power. These insights can be transferred to other economic structures. Another methodological limitation arises from the availability and accessibility of data. As emphasised by Hergesell, Hopkins and Wallerstein, mixed methods approaches and diverse databases are essential for longue durée research, but data collection is often laborious and not without limitations. During the time under study, it was therefore necessary to draw on different types of data and analyse them in relation to the research question. In some cases, I had to rely on pre-interpreted material due to limited access to historical sources and contemporary actors, as well as limited time for research. This meant comparing my own impressions with existing data. The book's empirical basis is largely based on process-produced data; with a longer research period, this could possibly have been supplemented by additional research-elicited data, particularly for period III. It should also be noted that, in terms of content, the processes and structures described are based on the identified and analysed data. It can not be ruled out that other relevant aspects may have been overlooked. Therefore, a complete study can not be guaranteed; however, theoretical saturation has resulted in a robust and differentiated picture.

Further research could involve conducting comparative case studies to analyse the long-term development of the cut flower industry in other regions. It would be particularly beneficial to compare regions that differ significantly from Aalsmeer, such as those that primarily produce for the domestic market (e.g. Kunming in China, Holambra in Brazil and Ontario in Canada), with regions that have been export-oriented from the outset (e.g. the Mount Kenya region in Kenya, Oromia in Ethiopia and Antioquia in Colombia).

Another area of research could focus on the spatial expansion of firms. Key questions here would include why firms choose certain locations, why production processes remain in one place or are relocated, how and for what reasons states support certain regions and how the interaction between states and firms is structured. While this book focused on processes of power establishment and maintenance, future work could more closely examine how external shocks, such as global crises or political upheavals, affect commodity chains (as suggested by Dannenberg et al.). Furthermore, regions that were formerly colonies but remain part of the European continent could be an interesting area for research. Santos and Boatcă refer to these regions as 'European Elsewheres'. French Guiana and the Falkland Islands are examples of such regions (Santos and Boatcă 2023). Moreover, it is necessary to systematically record the effects of the process-nature of commodity chains on social inequality, whether along the entire chain or at specific nodes, such as the actors or spatial structures. Both longue durée studies and focused analyses of one meaningful period could be suitable for this purpose.

A systematic comparison of case studies would be particularly desirable in all of the aforementioned research projects and within the longue durée, in order to increase the generalisability of the results and capture the diversity of economic structures in the modern world-system.

Over the last 300 years, cut flowers have evolved from objects of research and art to a commodity produced and traded globally.

The cut flower industry has deep roots in Dutch history and identity, as well as in processes such as imperialism, technological progress and globalisation. This book's analysis of the cut flower commodity chain, centred around the flower auction in Aalsmeer, highlights how this has laid the foundations for the Netherlands' 'flower nation' image to this day. Simultaneously, examining the longue durée reveals how the spatial expansion of the world-economy forms the basis of, and is an integral part of, processes of power establishment and maintenance. It is precisely this process-nature of the economic structure that enable actors to respond to internal and external processes.

References

Akremi, Leila. 2022. 'Stichprobenziehung in der qualitativen Sozialforschung'. In *Handbuch Methoden der empirischen Sozialforschung*, edited by Nina Baur and Jörg Blasius, 405–24. Wiesbaden: Springer Fachmedien. https://doi.org/10.1007/978-3-658-37985-8_26.

Alexander, R., Karen A. Gallant, Fenton Litwiller, Catherine White, and Barb Hamilton-Hinch. 2022. 'The Go-along Interview: A Valuable Tool for Leisure Research'. *Leisure Sciences* 42 (1): 51–68. https://doi.org/10.1080/01490400.2019.1578708.

Alford, Matthew, Reena das Nair, Margareet Visser, Stefano Ponte, and Shingie Chisoro. 2025. 'Due Diligence Regulation and Sustainability Governance in Value Chains: Lessons From the South African Wine Sector'. *Regulation & Governance*, May. https://doi.org/10.1111/rego.70038.

Algemeen Dagblad. 1995. 'Politiek Is Tevreden Met Schiphol, Bevolking Niet', 20 February 1995, Dag edition.

Algemeene Weekblad. 1857. 'Een Togtje in Den Haarlemmermeer-Polder', 15 April 1857.

Amīn, Samīr. 1974. *Accumulation on a World Scale : A Critique of the Theory of Underdevelopment*. Hassocks: Harvester.

Arendt, Hannah. 1970. *On Violence*. San Diego; New York; London: Harcourt Brace Jovanovich.

Arrighi, Giovanni. 1994. *The Long Twentieth Century: Money, Power and the Origins of Our Times*. London; New York: Verso Books.

Art Blossom. 2020. 'Economic Aspects of the Flower Production in the EU'. 1. https://ec.europa.eu/programmes/erasmus-plus/project-result-content/22063042-94e1-4dcf-89e6-380f72f50283/O1_ABLOOM.pdf.

Auctioneer. 2022. Site Visit with Narrative Interview: Area of Auctioneers: Interview by Lara Maria Espeter and Jakob Engel.

Bair, Jennifer. 2005. 'Global Capitalism and Commodity Chains: Looking Back, Going Forward'. *Competition & Change* 9 (2): 153–80. https://doi.org/10.1179/102452905X45382.

Bair, Jennifer, Mathew Mahutga, Marion Werner, and Liam Campling. 2021. 'Capitalist Crisis in the "Age of Global Value Chains"'. *Environment and Planning A: Economy and Space* 53 (6): 1253–72. https://doi.org/10.1177/0308518X211006718.

Bakker, Bas B. 1999. 'Crisis and Recovery'. In *The Netherlands: Transforming a Market Economy*, 4–15. Occasional Paper 181. Washington, DC: International Monetary Fund. https://doi.org/10.5089/9781557758040.084.

L. M. Espeter, *From Hobby to Billion-Dollar Industry*, Wirtschaft + Gesellschaft,
https://doi.org/10.1007/978-3-658-52437-1

Baraibar Norberg, Matilda and Lisa Deutsch. 2023. *The Soybean Through World History: Lessons for Sustainable Agrofood Systems.* Oxfordshire: Taylor & Francis. https://directory.doabooks.org/handle/20.500.12854/101577.

Bargawi, Hannah K., and Susan A. Newman. 2017. 'From Futures Markets to the Farm Gate: A Study of Price Formation along Tanzania's Coffee Commodity Chain'. *Economic Geography* 93 (2): 162–84.

Baur, Nina. 2005. *Verlaufsmusteranalyse: Methodologische Konsequenzen Der Zeitlichkeit Sozialen Handelns.* Wiesbaden: VS Verlag für Sozialwissenschaften.

Baur, Nina. 2017. 'Process-Oriented Micro-Macro-Analysis: Methodological Reflections on Elias and Bourdieu'. *Historical Social Research* 42 (4): 43–74.

Baur, Nina. 2019. 'Linearity vs. Circularity? On Some Common Misconceptions on the Differences in the Research Process in Qualitative and Quantitative Research'. *Frontiers in Education* 4 (June). https://doi.org/10.3389/feduc.2019.00053.

Baur, Nina. 2023. 'Long-Term Processes as Obstacles Against the Fourth Ecological Transformation: Ecological Sustainability and the Spatial Arrangements of Food Markets'. *Historical Social Research* 48 (1): 105–45.

Baur, Nina, and Gabriela Christmann. 2024. 'Case Selection and Generalization'. In *Handbook of Qualitative and Visual Methods in Spatial Research*, edited by Anna Juliane Heinrich, Séverine Marguin, Angela Million, and Jörg Stollmann, 1st ed., 77–90. Re-Figuration von Räumen. Bielefeld, Germany: transcript Verlag. https://doi.org/10.14361/9783839467343-007.

Baur, Nina, and Siegfried Lamneck. 2017. 'Einzelfallanalyse'. In *Qualitative Medienforschung. Ein Handbuch*, edited by Lothar Mikos and Claudia Wegener, 2., völlig überarbeitete und erweiterte Auflage, 290–300. Konstanz: UVK Verlagsgesellschaft mbH.

Beemster, Diana. 2017. 'The Future of Floriculture Logistics: A Logistic Design for the Floriculture Chain for the Year 2025, Incorporating Feasible Trends and Developments'. Delft, Niederlande.

Betuws(Ch) Tuinbouwblad. 1965a. 'Het Tuinbouwonderwijs in Het Rivierkleigebied', 19 June 1965.

Betuws(Ch) Tuinbouwblad. 1965b. 'Kwaliteit Is de Kracht van de Concurrentie', 1965.

Boatcă, Manuela. 2020. 'Du sollst den Spargel ehren'. KATAPULT-Magazin. 30 April 2020. https://katapult-magazin.de/de/artikel/du-sollst-den-spargel-ehren.

Bourdieu, Pierre. 1983. 'The Field of Cultural Production, or: The Economic World Reversed'. *Poetics* 12 (4): 311–56. https://doi.org/10.1016/0304-422X(83)90012-8.

Braudel, Fernand. 1996a. *The Mediterranean and the Mediterranean World in the Age of Philip II.* Vol. 1. Columbia: University Press Group.

Braudel, Fernand. 1996b. *The Mediterranean and the Mediterranean World in the Age of Philip II.* Vol. 2. Columbia: University of California Press.

Britannica. 2025. 'Iron Curtain'. 27 March 2025. https://www.britannica.com/event/Iron-Curtain.

burocontent. 'All You Need to Know about the World's Largest Flower Auction: Royal FloraHolland in Aalsmeer'. Visit Aalsmeer. 8 June 2021. https://www.visitaalsmeer.nl/en/facts-flower-auction-aalsmeer/.

Butollo, Florian, Cornelia Staritz, Felix Maile, and Tobias Wuttke. 2025. 'The End of Globalized Production? Supply-Chain Resilience, Technological Sovereignty, and Enduring

Global Interdependencies in the Post-Pandemic Era'. *Critical Sociology* 51 (4–5): 779–98. https://doi.org/10.1177/08969205241239872.
Centraal Bureau voor de Statistiek (CBS). "Population, Households and Population Dynamics; 1899–2019." StatLine. April 9, 2021. https://opendata.cbs.nl/statline/#/CBS/en/dataset/37556ENG/table.
Chase-Dunn, Christopher. 1998. *Global Formation: Structures of the World-Economy*. Lanham: Rowman & Littlefield Publishers, Inc.
Chief Inspector. 2022. Guided Interview: Governmental Organisation (Nairobi Airport) Interview by Jakob Engel and Lara Maria Espeter.
'Clusius Correspondence'. 2015. https://clusiuscorrespondence.huygens.knaw.nl/edition.
Coe, Neil M., Peter Dicken, and Martin Hess. 2008. 'Global Production Networks: Realizing the Potential'. *Journal of Economic Geography*, no. 8 (January), 271–95.
Dallas, Mark P., Stefano Ponte, and Timothy J. Sturgeon. 2019. 'Power in Global Value Chains'. *Review of International Political Economy* 26 (4): 666–94. https://doi.org/10.1080/09692290.2019.1608284.
Davies, Kasia. 2025a. 'Number of Cut Flowers Sold by Royal FloraHolland in 2020, by Type'. Statista. 2025. https://www.statista.com/statistics/829433/sales-volume-of-cut-flowers-by-royal-floraholland-by-type/.
Davies, Kasia. 2025b. 'Revenue of Cut Flowers Sold by Royal FloraHolland in 2022, by Type'. Statista. 2025. https://www.statista.com/statistics/829413/sales-value-of-cut-flowers-sold-by-royal-floraholland-by-type/.
De Dordrechtsche Courant. 1802, 14 September 1802.
De Dordrechtsche Courant. 1836, 29 December 1836.
De Dordrechtsche Courant. 1837, 8 December 1837.
De Dordrechtsche Courant. 1842, 22 November 1842.
De Dordrechtsche Courant. 1855, 8 September 1855.
De Zaanlander. 1940. 'Veilingomzetten Te Aalsmeer', 9 July 1940.
De Zaanlander. 1949. 'In Het Veilinggebouw. Bloemen Voor Binnen- En Buitenland', 28 May 1949.
Deventer Dagblad. 1923, 26 May 1923.
De Zaanlander. 1933. 'Erste Kamer. Landbouwsteunwetten', 2 May 1933.
Dobeson, Alexander. 2016. 'Hooked on Markets. Revaluing Coastal Fisheries in Liberal Rural Capitalism'. Uppsala University.
Dobeson, Alexander, and Sebastian Kohl. 2020. 'Why Durability Matters? Towards a Comparative Economic Sociology of Market Organization'. *European Journal of Sociology / Archives Européennes de Sociologie* 61 (1): 33–64. https://doi.org/10.1017/S0003975620000028.
'Dounan Flowers Market in Kunming'. n.d. Yaso Trip. Accessed 2 July 2025. https://www.yunnanexploration.com/dounan-flowers-market-in-kunming.html.
Engel, Jakob. 2025. *Die Wertschöpfungskette Kenianischer Rosen Für Den Deutschen Markt. Auswirkungen Exogener Schocks Auf Die Resilienz Globaler Warenketten*. Würzburg: Würzburg University Press.
Enkhuizer Courant. 1883. 'BINNENLAND', 20 June 1883.
Enkhuizer Courant. 1911. 'Land- En Tuinbouw', 11 March 1911.

Espeter, Lara M. 2025. 'Ohne Unten Kein Oben: Theoretische Grundlage Zur Betrachtung Globaler Wertschöpfungsprozesse Und Globaler Ungleichheiten Anhand Der Welt-System-Analyse Und Des Global Value Chain Ansatzes'. In *Wirtschaft Und Soziale Ungleichheit: Neue Soziologische Positionen*, edited by Nina Baur and Jürgen Beyer, 223–43. Wirtschaft + Gesellschaft. Wiesbaden: Springer VS. https://doi.org/10.1007/978-3-658-46565-0_11.

Espeter, Lara M. 2025. 'Power in Global Commodity Chains. Introducing Commodity Chain Mapping as a Method for Analysing Complex Spatialities'. *HSR – Historical Social Research, 50 (4)*, 344–377. https://doi.org/10.12759/hsr.50.2025.52.

Espeter, Lara M., and Linda Hering, eds. 2025. *Commodity Chains under Pressure. Resilience through Coping, Adaptation, and Transformation.* Cheltenham: Edward Elgar Publishing. https://doi.org/10.4337/9781035355303.

Espeter, Lara M., and Patricia Retamal. 2024. 'Global Commodity Chains and the Pandemic: Labor-Power in Agricultural Sectors in Kenya and Chile'. *Journal of World-Systems Research* 30 (1): 223–48. https://doi.org/10.5195/jwsr.2024.1193.

European Parliament. 2025. 'Treaty of Rome'. About Parliament. 2025. https://www.europarl.europa.eu/about-parliament/en/in-the-past/the-parliament-and-the-treaties/treaty-of-rome.

European Union. n.d.-a. "Euro – History and Purpose." European Union. Accessed July 11, 2025. https://european-union.europa.eu/institutions-law-budget/euro/history-and-purpose_en.

European Union.. n.d.-b. 'Greece – EU Country Profile'. Accessed 25 May 2025. https://european-union.europa.eu/principles-countries-history/eu-countries/greece_en.

European Union.. n.d.-c. 'Spain – EU Country Profile'. Accessed 25 May 2025. https://european-union.europa.eu/principles-countries-history/eu-countries/spain_en.

Farm Manager. 2022. Guided Interview: Cut Flower Farm (Naivasha) Interview by Jakob Engel and Lara Maria Espeter.

Fligstein, Neil. 2001. *The Architecture of Markets: An Economic Sociology of Twenty-First-Century Capitalist Societies.* Princeton: Princeton University Press.

'Flower Auction Japan'. n.d. Flower Auction Japan. Accessed 2 July 2025. https://www.faj.co.jp/en/.

Follmann, Alexander, Peter Dannenberg, Nina Baur, Boris Braun, Grit Walther, Amelie Bernzen, Jan Börner, et al. 2024. 'Conceptualizing Sustainability and Resilience in Value Chains in Times of Multiple Crises—Notes on Agri-Food Chains'. *DIE ERDE – Journal of the Geographical Society of Berlin* 155 (1): 29–48. https://doi.org/10.12854/erde-2024-692.

Foral Daily. 2023. 'Colombian Flower Industry in Numbers'. 11 October 2023. https://www.floraldaily.com/article/9567195/colombian-flower-industry-in-numbers/.

Former Manager. 2022. Guided Interview: Public-Benefit Organisation Interview by Jakob Engel and Lara Maria Espeter.

Frank, André Gunder. 1978. *World Accumulation : 1492 - 1789.* London: Macmillan.

Frederick, Stacey. 2021. 'Global Value Chain Mapping'. In *Handbook on Global Value Chains*, edited by Stefano Ponte, Gary Gereffi, and Gale Raj-Reichert, Paperback edition, 29–53. Cheltenham, UK; Northampton, MA, USA: Edward Elgar Publishing.

Freight Forwarder (focused KEN to NL). 2022a. Guided Interview: Freight Forwarder (focused KEN to NL) (Aalsmeer) Interview by Jakob Engel and Lara Maria Espeter.

Freight Forwarder (focused KEN to NL). 2022b. Guided Interview: Freight Forwarder (focused KEN to NL) (Nairobi Airport) Interview by Jakob Engel and Lara Maria Espeter.

Gasiorek, Michael, and Nicolò Tamberi. 2023. 'The Effects of Leaving the EU on the Geography of UK Trade'. *Economic Policy* 38 (116): 707–64.

Gebhardt, Andrew. 2014. *Holland Flowering. How the Dutch Flower Industry Conquered the World*. Amsterdam: Amsterdam University Press. https://doi.org/10.1515/9789048522590.

Gemählich, Andreas. 2022. *The Kenyan Cut Flower Industry and Global Market Dynamics*. Future Rural Africa. Woodbridge: James Currey.

Gereffi, Gary, John Humphrey, and Timothy Sturgeon. 2005. 'The Governance of Global Value Chain'. *Review of International Political Economy* 12 (1): 78–104.

Gereffi, Gary, David Spener, and Jennifer Bair, eds. 2002. *Free Trade And Uneven Development: The North American Apparel Industry After NAFTA*. Philadelphia: Temple University Press.

Goldgar, Anne. 2007. *Tulipmania. Money, Honor, and Knowledge in the Dutch Golden Age*. Chicago; London: University of California Press.

Goody, Jack. 1993. *The Culture of Flowers*. New York: Cambridge University Press.

Graeff, Peter, and Nina Baur. 2020. 'Digital Data, Administrative Data, and Survey Compared: Updating the Classical Toolbox for Assessing Data Quality of Big Data, Exemplified by the Generation of Corruption Data'. *Historical Social Research* 45 (3): 244–69. https://doi.org/10.12759/hsr.45.2020.3.244-269.

Griesi, Elettra. 2026. 'Entanglements Between the Refiguration of EU's Territorial Spaces and the Recruitment Manage-Ment of Seasonal Labour Migration'. *HSR – Historical Social Research*, 51 (1). https://doi.org/10.12759/hsr.51.2026.07.

Hale, Angela, and Maggie Opondo. 2005. 'Humanising the Cut Flower Chain: Confronting the Realities of Flower Production for Workers in Kenya'. *Antipode* 37 (2): 301–23.

Hartmann, Peter H., and Andrea Lengerer. 2022. 'Verwaltungsdaten und Daten der amtlichen Statistik'. In *Handbuch Methoden der empirischen Sozialforschung*, edited by Nina Baur and Jörg Blasius, 1537–45. Wiesbaden: Springer Fachmedien. https://doi.org/10.1007/978-3-658-37985-8_106.

Hayes, Charles. 1997. *Oserian: Place of Peace*. Nairobi: Rima Books.

Heck, Eric van. 2021. *Technology Meets Flowers. Unlocking the Circular and Digital Economy*. Cham: Springer.

Helfferich, Cornelia. 2022. 'Leitfaden- und Experteninterviews'. In *Handbuch Methoden der empirischen Sozialforschung*, edited by Nina Baur and Jörg Blasius, 875–92. Wiesbaden: Springer Fachmedien. https://doi.org/10.1007/978-3-658-37985-8_55.

Hergesell, Jannis. 2019. *Technische Assistenzen in Der Altenpflege: Eine Historisch-Soziologische Analyse Zu Den Ursachen Und Folgen von Pflegeinnovationen*. Weinheim: Beltz Juventa.

Hergesell, Jannis. 2021. 'Re-Figuration of Spaces as Long-Term Social Change: The Methodological Potential of Comparative Historical Sociology for Cross-Cultural Comparison'. *Forum: Qualitative Social Research* 22 (2).

Hergesell, Jannis, Nina Baur, and Lilli Braunisch. 2020. 'Process-Oriented Sampling'. *Canadian Review of Sociology – Revue Canadienne de Sociologie* 57 (2): 265–85. https://doi.org/10.1111/cars.12281.

Hering, Linda. 2022. *Die Materialität Der Märkte: Qualität, Bewertung Und Organisation Im Deutschen Obst- Und Gemüsehandel*. Re-Figuration von Räumen 4. Bielefeld: transcript.

Hering, Linda, and Lara M. Espeter. 2025. 'Globalised Agriculture and External Shocks: Theoretical and Methodological Implications on the Research of Commodity Chains'. In *Commodity Chains under Pressure. Resilience through Coping, Adaptation, and Transformation*, edited by Lara M. Espeter and Linda Hering. Cheltenham. https://doi.org/10.4337/9781035355303.00009.

Hering, Linda, and Robert Jungmann. 2022. 'Einzelfallanalyse'. In *Handbuch Methoden der empirischen Sozialforschung*, edited by Nina Baur and Jörg Blasius, 677–89. Wiesbaden: Springer Fachmedien. https://doi.org/10.1007/978-3-658-37985-8_42.

Het Nieuws. 1920. 'BINNENLAND', 9 July 1920.

Historische Tuin Aalsmeer. 2020. 'Historical Garden in Aalsmeer The Netherlands'. https://www.historischetuinaalsmeer.nl/wp-content/uploads/2020/12/HTAguide.pdf.

Hopkins, Terence K., and Immanuel Wallerstein. 1977. 'Patterns of Development of the Modern World-System'. *Review (Fernand Braudel Center)* 1 (2): 111–45.

Hopkins, Terence K., and Immanuel Wallerstein. 1986. 'Commodity Chains in the World-Economy Prior to 1800'. Review (Fernand Braudel Center), 10 (1): 157–170.

Hopkins, Terence K., and Immanuel Wallerstein. 1994a. 'Commodity Chains in the Capitalist World-Economy Prior to 1800'. In *Commodity Chains and Global Capitalism*, edited by Gary Gereffi and Miguel Korzeniewicz, 17–20. Westport, CT: Praeger. https://www.jstor.org/stable/40241052.

Hopkins, Terence K., and Immanuel Wallerstein. 1994b. 'Conclusions about Commodity Chains'. In *Commodity Chains and Global Capitalism*, edited by Gary Gereffi and Miguel Korzeniewicz, 48–50. Westport, CT: Praeger.

Hopkins, Terence K., and Immanuel Wallerstein. 2016. 'Structural Transformations of the World-Economy'. *Review (Fernand Braudel Center)* 39 (1/4): 171–94.

Huijs, Menno. 2011. 'Building Castles in the (Dutch) Air. Understanding the Policy Deadlock of Amsterdam Airport Schiphol 1989 - 200'. Delft: Delft University of Technology.

IMF. 2025 'GDP per Capita Netherlands from 1980 to 2030'. 2025. Statista. 18 June 2025. https://www.statista.com/statistics/263598/gross-domestic-product-gdp-per-capita-in-the-netherlands/.

International Trade Centre. 2025a. 'List of Exporters for the Selected Product in 2023 (Cut Flowers and Flower Buds of a Kind Suitable for Bouquets or for Ornamental Purposes, Fresh, Dried, Dyed, Bleached, Impregnated or Otherwise Prepared)'. Trade Map. 2025.

International Trade Centre. 2025b. 'Trade Map - List of Importers for the Selected Product in 2023 (Cut Flowers and Flower Buds of a Kind Suitable for Bouquets or for Ornamental Purposes, Fresh,...)'. Trade Map. 2025.

'Kaart van Den Ringdijk Tusschen Schiphol En Het Nieuwe-Meer, Mijlpalen 35–37 Waarop Met Kleuren Zijn'. 1867.

Karataşli, Şahan Savaş. i.P. 'Primitive Accumulations, Financial Expansions, and Varieties of Spatial Refiguration of the Capitalist World-Economy'. *HSR - Historical Social Research*, 50 (4): 410–444. https://doi.org/10.12759/hsr.50.2025.54.

Kennedy, James C. 2017. *A Concise History of the Netherlands*. Cambridge, UK: Cambridge University Press.

Kenya Flower Council. n.d. 'Our History'. Kenya Flower Council. Accessed 22 April 2025. https://kenyaflowercouncil.org/index.php/our-story/our-history.

Kirchner, Stefan, and Nina Baur. 2025. 'Spatially Bound Regimes between Convergence and Space Dependence. A Varieties-of-Refiguration Perspective on Social Change around the World'. *SFB 1265 Working Paper* 18 (April). https://doi.org/10.14279/DEPOSITONCE-23585.

Kondratiev, Nikolai Dmitriyevich. 1935. 'The Long Waves in Economic Life'. *The Review of Economics and Statistics* 17 (6): 105–15. https://doi.org/10.2307/1928486.

Korzeniewicz, Roberto Patricio, and Corey R. Payne. 2019. 'Sugar, Slavery, and Creative Destruction: World-Magnates and "Coreification" in the Longue-Durée'. *Journal of World-Systems Research* 25 (2): 395–419. https://doi.org/10.5195/jwsr.2019.893.

Kuiper, Gerda, and Clemens Greiner. 2021. 'Export Horticulture and Labour Migration in Kenya: Translocality and Transiency in a Secondary Town'. *Geoforum* 122 (January): 1–10. https://doi.org/10.1016/j.geoforum.2021.03.013.

'Kunming'. 2025. In *Wikipedia*. https://de.wikipedia.org/w/index.php?title=Kunming&oldid=254793303.

Küsters, Ivonne. 2022. 'Narratives Interview'. In *Handbuch Methoden der empirischen Sozialforschung*, edited by Nina Baur and Jörg Blasius, 893–900. Wiesbaden: Springer Fachmedien. https://doi.org/10.1007/978-3-658-37985-8_56.

'Largest Commercial Building'. n.d. Guinness World Records. Accessed 20 February 2025. https://www.guinnessworldrecords.com/world-records/69357-largest-commercial-building.html.

'Largest Flower Auction'. n.d. Guinness World Records. Accessed 20 February 2025. https://www.guinnessworldrecords.com/world-records/largest-flower-auction.html.

Lead Auditor. 2022. Guided Interview: Employer Organisation (Nairobi) Interview by Jakob Engel and Lara Maria Espeter.

Lee, Joonkoo, Gary Gereffi, and Sang-Hoo Lee. 2016. 'Social Upgrading in Mobile Phone GVCs. Firm-Level Comparisons of Working Conditions and Labour Rights'. In *Labour in Global Value Chains in Asia*, edited by Dev Nathan, Meenu Tewari, and Sandip Sarkar, 315–52. Cambridge: Cambridge University Press. https://doi.org/10.1017/9781316217382.

Li, Honglin, dir. 2024. "Exploring Flower Farms in Kunming, Yunnan: A Blossoming Journey." YouTube video, 19:38. 12 April 2024. https://www.youtube.com/watch?v=1VqMqXRif2o.

Lier, Bas van. 2005. *Het Groene Goud. Opkomst En Ontwikkeling van Bloemenveiling Aalsmeer*. Amsterdam: Meteor Press Licensing BV.

Logistic Expert Royal FloraHolland and Docservice Manager. 2022. Site Visit with Walking Interviews: Royal Flora Holland Interview by Jakob Engel and Lara Maria Espeter.

Maas- En Roerbode. 1947. 'Bloemen, Licht En Water in Lusthof Aalsmeer', 25 June 1947.

Mackay, Michael, Tracy Nelson, and Harvey C. Perkins. 2018. 'Interpretive Walks: Advancing the Use of Mobile Methods in the Study of Entrepreneurial Farm Tourism Settings'. *Geographical Research* 56 (2): 167–75. https://doi.org/10.1111/1745-5871.12275.

Managing Director. 2022. Narrative Interview: Supplier to the Cut Flower Industry Interview by Jakob Engel and Lara Maria Espeter.

Marx, Karl. 1887. *Capital: A Critical Analysis of Capitalist Production*. Moscow: Foreign Languages Publishing House.

Marx, Karl, and Friedrich Engels. 2015. *The Communist Manifesto*. Vol. no. 20. Penguin Classics. London: Penguin Classics.

Masselink, Maarten, and Paul van den Noord. 2009. 'The Global Financial Crisis and Its Effects on the Netherlands'. *ECFIN COUNTRY FOCUS* 6 (10). https://ec.europa.eu/economy_finance/publications/pages/publication16339_en.pdf.

Massey, Doreen. 1995. *Spatial Divisions of Labour: Social Structures and the Geography of Production.* London: Macmillan Education UK. https://doi.org/10.1007/978-1-349-24059-3.

Massey, Doreen, and Richard Meegan. 1982. *The Anatomy of Job Loss. The How, Why and Where of Employment Decline.* London; New York: Methuen.

Mekonnen, M. M., A. Y. Hoekstra, and R. Becht. 2012. 'Mitigating the Water Footprint of Export Cut Flowers from the Lake Naivasha Basin, Kenya'. *Water Resources Management* 26 (13): 3725–42. https://doi.org/10.1007/s11269-012-0099-9.

Meulenberg, Matthew T. G. 1989. 'Horticultural Auctions in the Netherlands: A Transition from "Price Discovery" Institution to "Marketing" Institution'. *Journal of International Food & Agribusiness Marketing* 1 (3): 139–65.

Ministerie van Klimaat en Groene Groei. n.d. 'Groningen Gasfield'. NLOG - Dutch Oil and Gas Portal. Accessed 30 June 2025. https://www.nlog.nl/en/groningen-gasfield.

Moore, Jason W. 2000. 'Sugar and the Expansion of the Early Modern World-Economy: Commodity Frontiers, Ecological Transformation, and Industrialization'. *Review (Fernand Braudel Center)* 23 (3): 409–33.

Nadri, Ghulam A. 2016. *The Political Economy of Indigo in India, 1580–1930. A Global Perspective.* European Expansion and Indigenous Response 22. Leiden; Boston: Brill.

Nationaalarchief. 2025a. 'Nederlandse Tuinbouwraad Te Den Haag: 1908'. Actorenregister. Nationaalarchief. https://actorenregister.nationaalarchief.nl/actor-organisatie/nederlandse-tuinbouwraad-te-den-haag.

Nationaalarchief. 2025b. 'Rijks Middelbare Tuinbouw School Te Aalsmeer: 1897'. Actorenregister. Nationaalarchief. https://actorenregister.nationaalarchief.nl/actor-organisatie/rijks-middelbare-tuinbouw-school-te-aalsmeer.

NDR. "die RECHERCHE: Inside Aldi: Folge 3: Die giftige Wahrheit der Billig-Rosen (S01/E03)." ARD Mediathek, September 4, 2023. https://www.ardmediathek.de/video/die-recherche-inside-aldi/folge-3-die-giftige-wahrheit-der-billig-rosen-s01-e03/ndr/Y3JpZDovL25kci5kZS80OTI0XzIwMjMtMDktMDQtMjEtMTU.

Netherlands Atlas. n.d. "Aalsmeer Municipality, North Holland." Netherlands Atlas. Accessed July 4, 2025. https://netherlandsatlas.com/North+Holland/Aalsmeer/.

'Netherlands - Trade Balance of Goods from 2013 to 2023'. 2025. Statista. 3 July 2025. https://www.statista.com/statistics/263627/trade-balance-of-goods-in-the-netherlands/.

Nieuw Noordhollandsch Dagblad. 1946. 'Gele Rozen En Viooltjes Voor Churchill', 14 May 1946.

NMa. 2007. 'Besluit van de Raad van Bestuur van de Nederlandse Ededingingsautoriteit Als Bedoeld in Artikel van de Mededingingswet. Bloemenveiling Aalsmeer - FloraHolland'. Openbare versie 5901/184. Nederlandse Mededingingsautoriteit.

Norkus, Maria, and Nina Baur. 2020. 'Time and Social Processes: Other Entries'. *SAGE Research Methods Foundations*, January. https://doi.org/10.4135/9781526421036.

Nowakowska, M. 2015. 'Safety of Supply Chain of Kenya's Cut Flower Industry. Case of Kongoni River Farm'. *Logistics and Transport* 2 (26): 43–48.

NRC Handelsblad. 1993. '"Als de Wind Tegenzit Gaan Ze, Bam, over de Stad"', 17 December 1993, Dag edition.

Nunspeets Nieuws- En Advertentieblad. 1982. 'Veiling Aalsmeer Nu Grootste Gebouw Ter Wereld', 17 March 1982.

'O Veiling'. n.d. *Veiling Holambra | Flores e Plantas Ornamentais* (blog). Accessed 2 July 2025. https://veiling.com.br/a-cooperativa/.

OEC. n.d. 'Cut Flowers in Colombia'. The Observatory of Economic Complexity. Accessed 20 February 2025. https://oec.world/en/profile/bilateral-product/cut-flowers/reporter/col.

OECD. 2014. *The Competitiveness of Global Port-Cities*. OECD. https://doi.org/10.1787/9789264205277-en.

Oliveira, Rogerio, Marlei Pozzebon, and Eric van Heck. 2007. 'Veiling Holambra - Trading Brazilian Flowers in the International Market Enabled by IT'. *Journal of Information Technology Case and Application Research* 9 (1): 38–58. https://doi.org/10.1080/15228053.2007.10856106.

Omosa, Mary, Mary Kimani, and Roseanne Njiru. 2006. 'The Social Impact of Codes of Practice in the Cut Flower Industry in Kenya'. Final Report. Nairobi: Natural Resources Institute, University of Nairobi.

O'Neill, Aaron. 2025a. 'Growth of the Real Gross Domestic Product (GDP) Netherlands from 1981 to 2030'. Statista. 18 June 2025. https://www.statista.com/statistics/263609/gross-domestic-product-gdp-growth-rate-in-the-netherlands/.

O'Neill, Aaron. 2025b. 'Netherlands - Distribution of Gross Domestic Product (GDP) across Economic Sectors 2013 to 2023'. Statista. 20 June 2025. https://www.statista.com/statistics/276713/distribution-of-gross-domestic-product-gdp-across-economic-sectors-in-the-netherlands/.

O'Neill, Aaron. 2025c. 'Netherlands - Distribution of the Workforce across Economic Sectors 2023'. Statista. 20 June 2025. https://www.statista.com/statistics/276716/distribution-of-the-workforce-across-economic-sectors-in-the-netherlands/.

Ontario Flower Growers. n.d. "Ontario Flower Growers." Accessed July 14, 2025. http://www.ontarioflowers.com.

Oostrom, Frits vaan. 2011. *The Netherlands in a Nutshell: Highlights from Dutch History and Culture*. Edited by Commissie Herijking Canon van Nederland. Amsterdam: Amsterdam Universiy Press. https://library.oapen.org/viewer/web/viewer.html?file=/bitstream/handle/20.500.12657/24322/9789048506149_optimised.pdf?sequence=1&isAllowed=y.

Owner. 2022a. Guided Interview: Cut Flower Farm (Naivasha) Interview by Jakob Engel and Lara Maria Espeter.

Owner. 2022b. Guided Interview: Import/Export Firm Interview by Jakob Engel and Lara Maria Espeter.

Özveren, Eyüp. 1994. 'The Shipbuilding Commodity Chain, 1590–1790'. In *Commodity Chains and Global Capitalism*, edited by Gary Gereffi and Miguel Korzeniewicz, 20–34. Westport, CT: Praeger.

Özveren, Y. Eyüp. 2000. 'Shipbuilding, 1590–1790'. *Review (Fernand Braudel Center)* 23 (1): 15–86. https://www.jstor.org/stable/40241478.

Paprika Filmes. Veiling_Institucional_Ingles. n.d. Accessed 2 July 2025. https://vimeo.com/paprikafilmes/review/330121205/22556aaffa.

Patel-Campillo, Anouk. 2010. 'Transforming Global Commodity Chains: Actor Strategies, Regulation, and Competitive Relations in the Dutch Cut Flower Sector'. *Economic Geography* 87 (1): 79–99. https://doi.org/10.1111/j.1944-8287.2010.01102.x.

Pelizzon, Sheila. 1994. 'The Grain Flour Commodity Chain, 1590–1790'. In *Commodity Chains and Global Capitalism*, edited by Gary Gereffi and Miguel Korzeniewicz, 34–47. Westport, CT: Praeger.

Pelizzon, Sheila. 2000. 'Grain Flour, 1590–1790'. *Review (Fernand Braudel Center)* 23 (1): 87–195. https://www.jstor.org/stable/40241479.

Plantion. n.d. 'About Us'. Plantion. Accessed 30 June 2025. https://www.plantion.nl/en/about-us/.

Ploeg, Ron van der. 2019. 'New Corporate Video Marks Holambra's 30th Anniversary'. *AIPH* (blog). 9 April 2019. https://aiph.org/floraculture/news/new-corporate-video-marks-holambras-30th-anniversary/.

Polanyi, Karl. 2005. *The Great Transformation. The Political and Economic Origins of Our Time*. Boston: Beacon Press.

Ponte, Stefano, Timothy J. Sturgeon, and Mark P. Dallas. 2021. 'Governance and Power in Global Value Chains'. In *Handbook on Global Value Chains*, edited by Stefano Ponte, Gary Gereffi, and Gale Raj-Reichert, Paperback edition, 120–37. Cheltenham, UK; Northampton, MA, USA: Edward Elgar Publishing.

Porter, Michael E., Jorge Ramirez-Vallejo, and Fred van Eenennaam. 2013. 'The Dutch Flower Cluster'. Harvard Business School: Case 711–507.

Prebisch, Raúl. 1963. *Towards a Dynamic Development Policy for Latin America*. New York: United Nations Publication.

Quark, Amy. 2014. 'Private Governance, Hegemonic Struggles, and Institutional Outcomes in the Transnational Cotton Commodity Chain'. In *The Political Economy of Commodity Chains*, edited by Jennifer Bair, 20:33–63. http://jwsr.pitt.edu/ojs/jwsr/article/view/572.

R., M. de. 1928. 'Hoe Anderen Ons Zien.' *Provinciale Noordhollandsche Courant*, 30 May 1928.

Rabobank, ed. 2022. 'World Floriculture Map 2021'.

Reilly, James A. 1993. 'From Workshops to Sweatshops: Damascus Textiles and the World-Economy in the Last Ottoman Century'. *Review (Fernand Braudel Center)* 16 (2): 199–213.

Research Working Group on Cyclical Rhythms and Secular Trends. 1979. 'Cyclical Rhythms and Secular Trends of the Capitalist World-Economy: Some Premises, Hypotheses, and Questions'. *Review (Fernand Braudel Center)* 2 (4): 483–500.

Responsible Official. 2022. Guided Interview: Governmental Organisation (Naivasha) Interview by Jakob Engel and Lara Maria Espeter.

Reuters. 2024. 'Groningen Gas Field in Netherlands to Shut down as Senate Approves Law', 16 April 2024, sec. Energy. https://www.reuters.com/business/energy/dutch-senate-approves-law-permanently-close-groningen-gas-field-2024-04-16/.

Riisgaard, Lone. 2009. 'Global Value Chains, Labor Organization and Private Social Standards: Lessons from East African Cut Flower Industries'. *World Development* 37 (2): 326–40. https://doi.org/10.1016/j.worlddev.2008.03.003.

Rong, Luo. 2022. 'China: A peek inside Kunming's Dounan Flower Market'. 23 February 2022. https://www.floraldaily.com/article/9403170/china-a-peek-inside-kunming-s-dounan-flower-market/.

Royal FloraHolland. 2022. 'Forward Together. Annual Report 2021'. 2022. Annual Report.

Royal FloraHolland. 2023. 'Together with You. 2023 Annual Report'.

Royal FloraHolland. n.d. 'Our Locations'. Royal FloraHolland. Accessed 30 June 2025. https://www.royalfloraholland.com/en/about-us/the-cooperative/our-locations.
Royal FloraHolland. n.d. 'Our History'. Accessed 13 May 2025. https://www.royalfloraholland.com/en/about-us/the-cooperative/our-history.
Salheiser, Axel. 2022. 'Natürliche Daten: Dokumente'. In *Handbuch Methoden der empirischen Sozialforschung*, edited by Nina Baur and Jörg Blasius, 1507–22. Wiesbaden: Springer Fachmedien. https://doi.org/10.1007/978-3-658-37985-8_104.
'Sanremo Flower Market'. n.d. *SanRemoGuide* (blog). Accessed 2 July 2025. https://sanremoguide.it/en/the-city-of-san-remo/san-remo-flowers/flower-market/.
Santos, Fabio, and Manuela Boatcă. 2023. 'European Elsewheres: Global Sociologies of Space and Europe'. In *Considering Space. A Critical Concept for the Social Sciences*, edited by Dominik Bartmanski, Henning Füller, Johanna Hoerning, and Gunter Weidenhaus. London: Routledge. https://doi.org/10.4324/9781003361152.
Sapoznik, Alexandra. 2019. 'Bees in the Medieval Economy: Religious Observance and the Production, Trade, and Consumption of Wax in England, c. 1300–1555'. *The Economic History Review* 72 (4): 1152–74.
Saxenian, AnnaLee. 2000. 'Regional Networks and Innovation in Silicon Valley and Route 128'. In *Regional Innovation, Knowledge and Global Change*. Routledge: London.
Schilder's Niewsblad. 1967. 'Bloemenliefhebbers Voor 't Laast Bijeen', 18 May 1967.
Schirmer, Julia. 2024. 'Beliebteste Schnittblumen in Deutschland nach Marktanteil in den Jahren 2022 und 2023'. Zentralverband Gartenbau. https://de.statista.com/statistik/daten/studie/206579/umfrage/beliebteste-schnittblumen-nach-marktanteil/.
Schirmer, Julia. 2025. 'Rosen: Wichtigste Lieferländer 2023'. Statista. 2025. https://de.statista.com/statistik/daten/studie/252806/umfrage/herkunftlaender-fuer-rosen/.
Schuitemakers Purmerender Courant. 1886. 'BINNENLAND', 30 May 1886.
Smith, Adam. 1999. *The Wealth of Nations*. London: Penguin Books.
Statista Research Department. 2024. 'Cut Flower Production by Province Canada 2023'. Statista. 2024. https://www.statista.com/statistics/476081/number-of-cut-flowers-produced-by-region-canada/.
Statista Research Department. 2025. 'Netherlands GDP per Capita by Province 2022'. Statista. 3 March 2025. https://www.statista.com/statistics/1279533/gdp-per-capita-netherlands-by-province/.
Statista Research Department. n.d. 'North Holland: GDP 2022'. Statista. Accessed 5 November 2024. https://www.statista.com/statistics/1402419/north-holland-gdp/.
'Statistieken gemeente Aalsmeer. Aantal inwoners per jaar'. 2016. AlleCijfers.nl. 1 July 2016. https://allecijfers.nl/gemeente/aalsmeer/#inwoners.
Steen, M. 2010. 'A World of Flowers: Dutch Flower Autions and the Market for Cut Flowers.' *Journal of Applied Horiculture 2010*, 1 January 2010. https://doi.org/10.4135/9781452229669.n742.
Stein, Daniel. 2004. *Die Wirtschaft Und Wirtschaftspolitik Der Niederlande*. München: GRIN Verlag.
Tabacková, Zuzana. 2024. 'Site Visits'. In *Handbook of Qualitative and Visual Methods in Spatial Research*, edited by Anna Juliane Heinrich, Séverine Marguin, Angela Million, and Jörg Stollmann, 1st ed., 7:199–212. Re-Figuration von Räumen. Bielefeld, Germany: transcript Verlag. https://doi.org/10.14361/9783839467343-017.

Talbot, John M. 2004. *Grounds for Agreement: The Political Economy of the Coffee Commodity Chain*. Oxford: Rowman & Littlefield Publishers Inc.

Talbot, John M. 2011a. 'The Coffee Commodity Chain in the World-Economy: Arrighi's Systemic Cycles and Braudel's Layers of Analysis'. *Journal of World-Systems Research*, February, 58–88. https://doi.org/10.5195/jwsr.2011.427.

Talbot, John M. 2011b. 'The Coffee Commodity Chain in the World-Economy: Arrighi's Systemic Cycles and Braudel's Layers of Analysis'. *Journal of World-Systems Research* 17 (1): 58–88.

Taurel, Charles Edouard. 1867. *Haarlemmermeer, 1667–1867*. Grafiek Drukwerk. https://proxy.archieven.nl/0/E63B2660FB8E11DF9E4D523BC2E286E2.

The Observatory of Economic Complexity. n.d.-a. "Cut Flowers." The Observatory of Economic Complexity. Accessed July 11, 2025. https://oec.world/en/profile/hs/cut-flowers.

The Observatory of Economic Complexity. n.d.-b. "Cut Flowers in Chinese Taipei." The Observatory of Economic Complexity. Accessed July 2, 2025. https://oec.world/en/profile/bilateral-product/cut-flowers/reporter/twn.

The Observatory of Economic Complexity. n.d.-c. "Cut Flowers in Netherlands Trade." The Observatory of Economic Complexity. Accessed July 3, 2025. https://oec.world/en/profile/bilateral-product/cut-flowers/reporter/nld.

Thierbach, Cornelia. 2024. 'How to Use Guided Interviews in Spatial Research'. In *Handbook of Qualitative and Visual Methods in Spatial Research*, edited by Anna Juliane Heinrich, Séverine Marguin, Angela Million, and Jörg Stollmann, 1st ed., 7:131–40. Re-Figuration von Räumen. Bielefeld, Germany: transcript Verlag. https://doi.org/10.14361/9783839467343-011.

Trade Union Official (NL, EU, International). 2022. Guided Interview: Trade Union (NL) Interview by Jakob Engel and Lara Maria Espeter.

'UFG Customer Access: About the Auction'. 2025. United Flower Growers. 2025. https://www.ufgca.com/store/about-the-auction.php.

UNESCO World Heritage Centre. n.d. 'Droogmakerij de Beemster (Beemster Polder)'. UNESCO World Heritage Centre. Accessed 8 May 2025. https://whc.unesco.org/en/list/899/.

Upper Management. 2022a. Guided Interview: Employer Organisation (Naivasha) Interview by Jakob Engel and Lara Maria Espeter.

Upper Management. 2022b. Site Visit with walking Interview: Industrialised Flower Farm Interview by Jakob Engel and Lara Maria Espeter.

Upper Management. 2022c. Site Visit with walking Interview: Open-Field Flower Farm Interview by Jakob Engel and Lara Maria Espeter.

'US: Best Selling Cut Flowers 2019'. n.d. Statista. Accessed 8 November 2024. https://www.statista.com/statistics/1300259/best-selling-cut-flowers-us/.

United Flower Growers. n.d. "About Us." United Flower Growers. Accessed July 2, 2025. https://ufg.co.nz/about.

Utrechtsch Nieuwsblad. 1933a. 'De Steun Aan Land- En Tuinbouw. Voorloopig Verslag Der Tweede Kamer', 4 April 1933. https://proxy.archieven.nl/0/FE3BE24D9F884A31877DD27923F109E1.

Utrechtsch Nieuwsblad. 1933b. 'Landelijke Organisatie van Bloemenveilingen', 31 May 1933, 27000 edition.

Utrechtsch Nieuwsblad. 1951. 'Kweekster', 19 May 1951.

Vellekoop, Elita. 2024. 'US: Cut Flower Ocean Shipments Continue to Surge'. Floral Daily. 15 February 2024. https://www.floraldaily.com/article/9600488/us-cut-flower-ocean-shipments-continue-to-surge/.

Vereniging Bloemenveiling Nederland. n.d. 'VBN Activities'. Vereniging Bloemenveiling Nederland. Accessed 30 June 2025. https://www.vbn.nl/en/vbn-activities/.

Vieira, Pedro Antonio. 2014. '"Brazil" in the Capitalist World-Economy from 1550 to c. 1800: An Empirical Demonstration through the Sugar Commodity Chain'. *Review (Fernand Braudel Center)* 37 (1): 1–34.

Vogl, Susanne. 2014. 'Gruppendiskussion'. In *Handbuch Methoden der empirischen Sozialforschung*, edited by Nina Baur and Jörg Blasius, 581–86. Wiesbaden: Springer Fachmedien. https://doi.org/10.1007/978-3-531-18939-0_41.

Wallerstein, Immanuel. 1989. *The Modern World-System III: The Second Era of Great Expansion of the Capitalist World-Economy, 1730s-1840s.* San Diego: Academic Press.

Wallerstein, Immanuel. 2000a. 'Introduction'. *Review (Fernand Braudel Center)* 23 (1): 1–13.

Wallerstein, Immanuel. 2000b. 'Next Steps'. *Review (Fernand Braudel Center)* 23 (1): 197–99.

Wallerstein, Immanuel. 2006. *World-System Analysis: An Introduction.* 4th ed. Durham and London: Duke University Press.

Wallerstein, Immanuel. 2011a. *The Modern World-System I: Capitalist Agriculture and the Origins of the European World-Economy in the Sixteenth Century.* Berkeley; Los Angeles; London: University of California Press.

Wallerstein, Immanuel. 2011b. *The Modern World-System II: Mercantilism and the Consolidation of the European World-Economy, 1600–1750.* Berkeley; Los Angeles; London: University of California Press.

Wallerstein, Immanuel. 2011c. *The Modern World-System IV: Centrist Liberalism Triumphant, 1789–1914.* Berkeley: University of California Press.

Weber, Max. 2019. *Economy and Society. An Outline of Interpretive Sociology.* Edited by Guenther Roth and Claus Wittich. Cambridge, Massachusetts: Harvard University Press. https://doi.org/10.4159/9780674240827.

'Welcome to United Flower Growers'. 2025. United Flower Growers. 2025. https://www.ufgca.com/.

Wiersinga, Rolien, Tycho Vermeule, Joost Snels, Roberto Fray da Silva, and Siert Wiersema. 2013. 'Sustainable Vegetable Chain Development in Brazil. Grape Tomatoes from Holambra to the Higher Market Segment'. *Wageningen University & Research Publications* Rapport GTB-1262.

Windeler, Arnold. 2001. *Unternehmungsnetzwerke. Konstitution Und Strukturation.* Wiesbaden: Westdeutscher Verlag.

Wintle, Michael. 2000. *An Economic and Social History of the Netherlands, 1800–1920. Demographic, Economic and Social Transition.* Cambridge, UK: Cambridge University Press.

Yin, Robert K. 2018. *Case Study Research and Applications: Design and Methods.* Sixth Edition. Los Angeles ; London ; New Delhi ; Singapore ; Washington DC ; Melbourne: SAGE.

Zanden, J.L. van. 1987. *De Economische Ontwikkeling van de Nederlandse Landbouw in de Negentiende Eeuw 1800–1914*. Utrecht: Landbouwhogeschool. https://doi.org/10.18352/bmgn-lchr.2796.

Zandvoort, Ton van, dir. 2009. *A Blooming Business: A Poetic Documentary About People Who Are Imprisoned by the Power of the Global Flower Industry.*

Ziegler, Catherine. 2007. *Favored Flowers. Culture and Economy in a Global System*. New York: Duke University Press. https://doi.org/10.1515/9780822390015.

'市場沿革'. n.d. Taipei Flower Auction CO., LTD. Accessed 2 July 2025. https://www.tflower.com.tw/TFAWebUI/WebPages/Index/1.

GPSR Compliance
The European Union's (EU) General Product Safety Regulation (GPSR) is a set of rules that requires consumer products to be safe and our obligations to ensure this.

If you have any concerns about our products, you can contact us on

ProductSafety@springernature.com

In case Publisher is established outside the EU, the EU authorized representative is:

Springer Nature Customer Service Center GmbH
Europaplatz 3
69115 Heidelberg, Germany

www.ingramcontent.com/pod-product-compliance
Ingram Content Group UK Ltd.
Pitfield, Milton Keynes, MK11 3LW, UK
UKHW040805101026
476460UK00001B/8
* 9 7 8 3 6 5 8 5 2 4 3 6 4 *